Padmanabhan Krishnan, Sharan Chandran M
Self-Reinforced Polymer Composites

Also of Interest

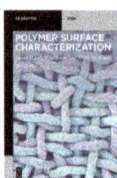

Polymer Surface Characterization
Luigia Sabbatini, Elvira De Giglio (Eds.), 2020
ISBN 978-3-11-070104-3, e-ISBN (PDF) 978-3-11-070109-8

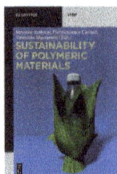

Sustainability of Polymeric Materials
Valentina Marturano, Veronica Ambrogi, Pierfrancesco Cerruti (Eds.),
2020
ISBN 978-3-11-059093-7, e-ISBN (PDF) 978-3-11-059058-6

Handbook of Biodegradable Polymers
Catia Bastioli (Ed.), 2020
ISBN 978-1-5015-1921-5, e-ISBN (PDF) 978-1-5015-1196-7

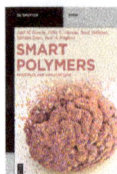

Smart Polymers.
Principles and Applications
José Miguel García, Félix Clemente García, José Antonio Reglero Ruiz,
Saúl Vallejos, Miriam Trigo-López, to be published in 2022
ISBN 978-1-5015-2240-6, e-ISBN (PDF) 978-1-5015-2246-8

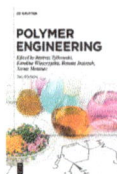

Polymer Engineering
Bartosz Tylkowski, Karolina Wieszczycka, Renata Jastrząb, Xavier
Montane (Eds.), to be published in 2022
ISBN 978-3-11-073844-5, e-ISBN (PDF) 978-3-11-073382-2

Padmanabhan Krishnan, Sharan Chandran M

Self-Reinforced Polymer Composites

The Science, Engineering and Technology

DE GRUYTER

Authors
Prof. Padmanabhan Krishnan
VIT University
Near Katpadi Road
Vellore 632014
Tamil Nadu
India
padmanabhan.k@vit.ac.in

Asst. Prof. Sharan Chandran M
VIT University
Near Katpadi Road
Vellore 632014
Tamil Nadu
India
sharanchandran.m@vit.ac.in

ISBN 978-3-11-064729-7
e-ISBN (PDF) 978-3-11-064733-4
e-ISBN (EPUB) 978-3-11-064740-2

Library of Congress Control Number: 2022933751

Bibliographic information published by the Deutsche Nationalbibliothek
The Deutsche Nationalbibliothek lists this publication in the Deutsche Nationalbibliografie;
detailed bibliographic data are available on the Internet at http://dnb.dnb.de.

© 2022 Walter de Gruyter GmbH, Berlin/Boston
Cover image: weisschr/iStock/Getty Images Plus
Typesetting: Integra Software Services Pvt. Ltd.
Printing and binding: CPI books GmbH, Leck

www.degruyter.com

We deeply acknowledge

The VIT management headed by Honorable Chancellor Dr Viswanathan G

Dean Dr Devendranath Ramkumar, former deans Dr Vasudevan R and Dr Arivazhagan N, and all heads of the departments of School of Mechanical Engineering, VIT

Mrs. S Kalaivani and other staff members of Central library, VIT Vellore

Mr Xavier of AMPT lab, staff of machine shop, and Ramya M, who was project associate of Dr. Padmanabhan K

SEM lab, SBST, VIT

Analytical chemistry lab in charge and supporting staff for DSC, TGA, and FTIR, VIT

Mr Aswin Suresh, former MSc chemistry student, VIT Vellore

Mr Gopinath K V, research scholar and JRF, VIT Vellore

Mr Ben Alex Baby, PhD research scholar, University of Palermo, Italy (former PG student of SMEC, VIT Vellore)

Mr Dipin Raj D K, senior consultant, Capgemini Technology Services India Limited (former PG student of SMEC, VIT Vellore)

Mr Kartic Hari Narayanan, former UG student, VIT Vellore

Dr. Sreethul Das, SMEC, VIT Vellore

Our family members, friends, and colleagues, and De Gruyter for having come forward to publish our book, which is the first ever book on self-reinforced composites

https://doi.org/10.1515/9783110647334-202

Contents

Chapter 1
Introduction

1.1 Overview

The transportation industry is one of the most dynamic sectors in the world and is always pressurized to reduce fuel consumption and waste accumulation. Durable and recyclable parts, with improved fuel efficiency, are the strategic aim of transportation companies for the last few decades. Self-reinforced composites (SRC) are one of the milestones in the development of such materials. Both the matrix and the reinforced materials of SRCs are composed of materials from the same family. One such popular material developed in recent history is the carbon/carbon composite. Still, nowadays the term "SRC" has been widely used to streak polymer-based SRCs. SRCs are also referred to as one polymer composite, single polymer composite, all-polymer composite, or homopolymer composite in the review literature [1–5].

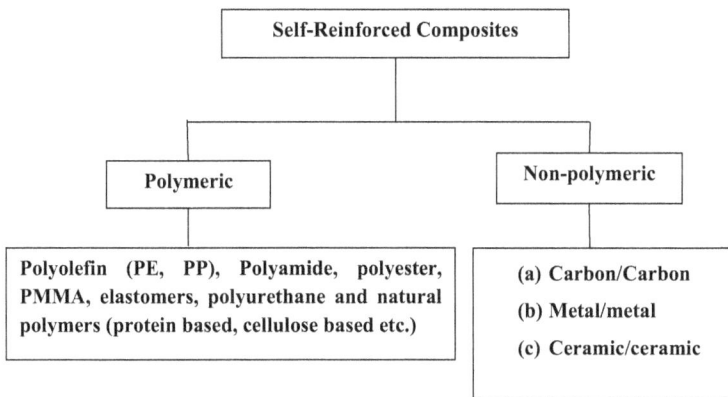

```
                    ┌─────────────────────────────┐
                    │  Self-Reinforced Composites │
                    └─────────────────────────────┘
                                  │
             ┌────────────────────┴────────────────────┐
     ┌───────────────┐                        ┌──────────────────┐
     │   Polymeric   │                        │   Non-polymeric  │
     └───────────────┘                        └──────────────────┘
┌────────────────────────────────┐   ┌──────────────────────────────┐
│ Polyolefin (PE, PP), Polyamide,│   │  (a) Carbon/Carbon           │
│ polyester, PMMA, elastomers,   │   │  (b) Metal/metal             │
│ polyurethane and natural       │   │  (c) Ceramic/ceramic         │
│ polymers (protein based,       │   │                              │
│ cellulose based etc.)          │   │                              │
└────────────────────────────────┘   └──────────────────────────────┘
```

Figure 1.1: Classification of self-reinforced composites.

SRCs can be widely classified as polymeric types and nonpolymeric types as shown in Figure 1.1. Natural SRCs can be observed in nature. Wood cellulose and animal muscle are examples of natural SRCs. Some of the SRCs developed from renewable resources are used as biomaterials.

Self-reinforced polymer composites (SRPCs) can be beneficial to the industries that are dedicated to working for reduced impact on the environment. They possess improved chemical functionality compared to other traditional composites due to identical chemical structures. Lightweight materials are gaining attention, especially in air cargo, shipping, and transportation. SRPCs can be completely recycled by

https://doi.org/10.1515/9783110647334-001

melting and result in zero waste accumulation. They indirectly help in gaining carbon credits by cutting the emission of carbon or greenhouse gases to the atmosphere.

Europe is one of the major manufacturers of composite materials. Production volume of glass fibre-reinforced plastics is observed to be at its highest level in 2019. This is an example of an increase in plastic waste accumulated in the environment.

> The **European parliament** has voted to ban **single-use plastic** cutlery, cotton buds, straws and stirrers as part of a sweeping law against **plastic** waste that despoils beaches and pollutes oceans. The vote by MEPs paves the way for a ban on **single-use plastics** to come into force by 2021 in all **EU** member states. EU member states will have to introduce measures to reduce the use of plastic food containers and plastic lids for hot drinks. By 2025, plastic bottles should be made of 25% recycled content, and by 2029 90% of them should be recycled.
>
> (*The Guardian*, 27 March 2019)

Similar regulations were implemented around the globe in the last few years. On 23 September 2019, India's prime minister had assured the United Nations that India will lead the world in efforts to crap single-use polymers. India directly implemented BS-VI (Bharat stage VI – an emission control measure for automobiles) since April 2020 skipping stage V of Indian regulations for controlling emission of greenhouse gases from automobiles though this decision caused an economic slowdown in India a few months before COVID-19 lockdown. Concern over weight reduction to meet the tight environmental guidelines over fuel efficiency of aviation and automobile products was the trigger of the development of SRCs, recyclability being an added advantage. Reinforcing a matrix material with a lighter material may result in a composite material with weak properties. Reinforcing a material at least of the same family or compound will give rise to a better material without compromising other properties. The same family of materials is also expected to possess better chemical compatibility and hence an improved interfacial bonding.

Ceramic/ceramic SRCs and metal/metal SRCs are to be treated differently from polymer SRCs. It is hard to find out any noticeable work done in pure SRCs from ceramics or metals. They may be prepared from the compound of the same element and this similarity is helping in reinforcement. This reason may be taken as the criteria to classify them as SRCs. But it is more logical to classify them as pseudo-SRCs.

1.2 Ceramic/ceramic SRC

There are several composite materials composed of a common source of ceramic materials. They can be termed as ceramic self-reinforced materials. This session introduces some of the recent developments in ceramic SRCs which would be useful in gaining a brief idea about the family of SRCs. The development of such SRCs gained momentum just two decades ago. They may not contribute to the appreciable properties of SRPC as they are not recyclable or weight gainers. Still, some of the properties can be attributed to them by keeping SRPC as referent SRCs.

Ceramic SRCs differ from polymer SRCs right from the beginning of the processing procedure. Self-reinforcement in ceramic SRCs may be obtained by forming a multiphase microstructure with one of the phases forming the matrix and the second one as the reinforcement. It is also possible by heat treatment in which one of the phases will be precipitated to form the matrix phase. Growing elongated intertwined grains will also provide a self-reinforcement in some cases. Careful control of temperature plays an important role in ceramic SRCs just like in polymer SRCs. Additionally, control over the formation of microstructure becomes another characteristic feature in ceramic SRCs.

Silicon nitride ceramics are chemically inert and possess excellent high-temperature strength, low coefficient of thermal expansion, oxidation resistance, high creep resistance, and thermal shock resistance compared to other high-temperature structural materials. Still, covalent bonds in those materials make them difficult to sinter and increase manufacturing costs. It is not easy to fabricate complex-shaped components because of the covalent bond. Gas pressure sintering or hot pressing are used to dense silicon nitride ceramics. Pressureless sintering is a method to overcome the issues in manufacturing. In situ self-reinforced textured microstructure can be produced by a combination of seeding, anisotropic grain growth and shear forming processes [6].

Although the preparation of silicon nitride was first reported in 1857, self-reinforcement researches in silicon nitride were developed only in the last two decades. Si_3N_4 is used in many applications like auto turbochargers, diesel engines, hydraulic pumps, bearings, spinal fusion devices replacing titanium, and polyether ether ketone. Matrix material like $Ba_2Al_2Si_2O_8$ (barium aluminosilicate, BAS) is poor in mechanical properties, and structural applications are limited. When BAS was reinforced with silicon nitride (Si_3N_4) by various processes like sintering, the resulting composite was a pseudo-SRC of silicon [7]. A promising class of materials with various structural applications including aerospace are being developed.

1.3 Metal/metal SRC

There is no universally accepted definition for a composite material. While defining and classifying SRCs, the same dilemma exists. However, the materials are classified as composites based on their load transferability from matrix to the reinforcement. Reinforcement need not be a material added from outside as we have seen in previous examples. Sometimes, the load-carrying ability of metals is improved by hindering the dislocation motion with small particles by dispersion or precipitation.

Pseudo-SRCs of discontinuously reinforced titanium matrix composites are gaining attention at a faster pace due to their wide acceptance in the automobile, aerospace, and aviation industries. They possess high specific strength, high-temperature resistance properties, and corrosion resistance. Thermal expansion coefficient is an important property to be considered for the materials used in high-temperature applications. Titanium boride (TiB) whisker is one of the suitable reinforcements in this regard for titanium matrices as it also possesses high elastic modulus and better interfacial bonding with titanium matrix. In TiB-reinforced Ti, an interfacial layer of TiB was formed by the precipitation of boron into the matrix as needles on the grain boundaries [8]. Another example in titanium-based composites is the development of research in coatings of titanium composites. TiB and TiC particles reinforced on Ti_6Al_4V by laser cladding using $Ti-B_4C-Al$ or $Ti-B_4C-C-Al$ powders as the precursor materials also fall under the category of SRCs [9].

1.4 Carbon/carbon SRC or all carbon composites

Carbon fibres were used by Thomas Edison as filaments of light bulbs in 1800. Industrial development of carbon fibres began in 1886 when National Carbon Company was established in Cleveland, Ohio [10]. Properties of carbon fibres improved later over two centuries and these fibres became one of the key materials in high-temperature, low coefficient of thermal expansion applications. Composite materials developed with carbon fibres as reinforcement offered a wide range of application options with improved mechanical properties [11].

Carbon exists in a variety of forms like graphite, diamond, graphene, fullerene, and carbon nanotubes. Graphite and fullerene are the allotropic forms of carbon widely used in carbon/carbon composites. The major limitation of carbon is its reaction with oxygen and the release of gaseous oxides of carbon. These composites were developed for space applications and jet vanes, used in German V_2 rockets in the beginning and later has got wide acceptance. They are also used in brake disk systems in racing cars, aircraft brake systems, and so on.

There are various processing routes for carbon/carbon composites. Sintering is not a possible way of processing these types of composites. Carbon matrix in a carbon/carbon composite is obtained generally by chemical vapour deposition of carbon or thermal decomposition of pitch or phenolic resin. However, the resulting composite will not possess the required density and strength due to the presence of pores.

High-pressure impregnation carbonization is one of the methods to fabricate carbon/carbon composites. A woven carbon fabric is impregnated with thermoplastic pitch under heat and pressure followed by pyrolysis of the pitch into carbon. Polymer matrix composites reinforced with carbon fibres can be converted to carbon matrix composites by converting the polymer into carbon by pyrolysis. In liquid-phase impregnation process, coal/tar/petroleum pitches and high carbon-yielding pitches are impregnated in the carbon fibre architecture.

Mechanical properties like strength-to-weight ratio and specific stiffness are high in carbon/carbon composites that are five times lighter than steel and two times lighter than aluminium. They are stable even above 3,000 °C with negligible thermal expansion when coated with thermal materials. They offer excellent fatigue properties and resistance to wear under hazardous conditions and high temperatures.

A major limitation of the carbon––carbon composites is their lack of impact resistance. Catastrophic failure of the space shuttle *Columbia* during atmospheric reentry was observed to be due to the breaking of one of its reinforced carbon–carbon panels by the impact of a piece of foam insulation from the space shuttle external tank [12].

1.5 Self-reinforced polymer composites

Polymers can be classified as thermoplastics and thermosetting plastics based on their molecular bonding. Weak Van der Waals forces in the thermoplastics can be broken down by heat while the thermosetting plastics possess cross-link bondings which make them one-time heat-processable materials. Thus, they disqualify themselves for processing as SRPC, considering the most favourable advantage of SRPC is recyclability. Thermoplastics can be processed by various methods like injection moulding, blow moulding, and extrusion moulding as they are more ductile compared to thermosetting polymers.

Thermoplastic polymers possess a different range of crystallinity. Semicrystalline nature of polymers exhibits a wide range of melting temperatures, which makes them suitable for thermal processing techniques in manufacturing. Semicrystalline polymers are widely used for reinforcement due to their higher stiffness and strength than amorphous ones. Semicrystalline or amorphous polymer grades can also be used for matrices. These materials have comparatively low density and costless, but lack in mechanical properties. Reinforcements like glass fibre, carbon, or aramid improve their properties by compromising on density, cost, reusability, ease of manufacturing, and so on. SRPCs are better candidates in this regard.

Fibres, tapes, or particles of the thermoplastics were used as reinforcements to make SRCs so far, but they differ from traditional concepts of composites in various aspects. SRCs are made from the same family of materials but may differ slightly in some of their properties due to changes in molecular weight (e.g. polyethylene (PE)), stereochemistry (e.g. polypropylene (PP)), and so on. These differences in properties impart the difference in melting temperatures of the matrix and reinforcement providing a proper window for processing. The chemical structure of matrix and reinforcement does not change, unlike the conventional composite materials. Similarities in the chemical structure of constituent materials provide better wetting and interfacial properties to the composites due to interdiffusion.

Mechanical properties of polymers are sensitive to temperature changes below the glass transition temperature (T_g). T_g is the temperature below which the polymers become hard and brittle like glass. The temperature and modulus of polymers are inversely proportional. As the difference in melting temperature is the major criterion while selecting the constituent matrix and reinforcement in SRPCs, a proper processing window has to be selected to utilize the strength and stiffness of the reinforcement completely. While the matrix melts, the reinforcement should remain intact. Degradation of the reinforcing material deteriorates the properties of the resulting composite. Selecting proper constituent materials available in the market from the same family of polymers with a workable temperature processing window is one of the challenges faced during the processing of SRPCs.

Molecular weight also plays an important role in the case of thermoplastics. If the molecular weight is higher, the polymer becomes a viscoelastic fluid at about 100 °C above T_g. At this temperature, the polymer can be treated as a melt. Other factors that determine the polymer's acceptance for processing as SRC are (i) percentage of crystallinity and (ii) melt viscosity. The modulus increases with the increasing crystallinity. Higher melt viscosity causes reduced matrix flow. Higher axial strength and stiffness result in fibres becoming anisotropic due to weak Van der Waals forces and hydrogen bond formation perpendicular to the molecular chains. Transverse strength and the compressive strength reduce. In the melt impregnation technique, the temperature window for processing the composite without causing damage to the fibres is very narrow, and the melt viscosity of the polyethylene (PE) matrix in this processing window is relatively high. Methods like hot compaction, overheating, film stacking, partial solution dissolving, cold drawing, and physical and chemical treatments are also developed for the processing of SRCs.

1.6 Major features of SRPCs

Some of the major observations from the literature are summarized here:
a) SRPCs are useful in making lightweight structures
b) They have the capability of retention of properties even at higher temperatures
c) They can be recycled
d) They possess a better interface

Some of the key properties of SRPCs recorded in literature are
a) Elevation of the melting point due to constrained orientation of fibres while processing.
b) Effect of compaction temperature and pressure: PP is more amorphous than PE. Relatively strong intermolecular forces in semicrystalline polymers prevent softening even above the glass transition temperature. Thus, PP is more sensitive to compaction temperature. A slight change in the compaction temperature affects

the properties. But PE is very stable even a few degrees below the melting point. Compaction pressure is another important parameter. Compaction pressure supports in the transfer of heat from the furnace to matrix–reinforcement assembly confirms proper joining of matrix and reinforcement and reduces the shrinkage effects of fibres or tapes.

c) PE fibre surfaces melt during heat treatment and form lamellae after crystallization. This can be observed as shish-kebab structure and epitaxial crystallization.

d) Comparable thermal expansion is an important advantage in SRCs. Thermal mismatch is minimum in SRPCs as they are developed from the same family of polymers. Still, molecular structure, fabric structure, and orientation, and the elastic anisotropywill influence the thermal expansion. Thus, the final molecular structure after hot compaction is the key factor in thermal expansion rather than the initial structure.

e) Transcrystallization, which is a phenomenon of nucleation of crystallization along the interface usually occurring in thermoplastics, observed in SRPCs seems to be a function of a combination of the factors like the topography of fibre, epitaxy of fibre and matrix, chemical composition of the fibre surface, thermal conductivity of the fibre, surface energy of the fibre, crystallinity of matrix, and processing conditions like cooling rate and hot compaction temperature.

1.7 Economic background

The global economic recession was a great challenge for the automotive industry from 2008 and many companies shut down their manufacturing units. However, since 2010 situation gradually changed and a boom in the automotive sector was observed. A bigger challenge faced by the automotive companies was to find a solution for the tight environmental regulations without compromising the increasing needs of performance by the customers. To meet the increased efficiency and emissions standards, automotive companies have to reduce the weight of the vehicles. SRPCs deliver considerable weight savings to the automobile industry.

To meet the requirements of climate change guidelines, more reusable lightweight materials should be used in automobiles, aircraft, and marine applications. Recyclable SRCs without compromising on the properties will be a better replacement. Curv® is one of the self-reinforced PP available in the market with a high impact performance at a temperature of −30 °C, and it remains ductile even at very low temperature. The density of Curv® is 920 kg/m^3 while that of PP is 900 kg/m^3 which indicates that the density has not increased much when the PP is reinforced unlike the glass fibre or carbon-reinforced PP. Notched Izod impact strength at 20 °C shows 100 times improvement in energy absorption during impact (400 kJ/m^2). It also shows five times better tensile strength, and the tensile modulus is 4.2 GPa while that of PP is just 1.12 GPa.

Curv®/PP honeycomb laminates and Curv®/PP foam laminates with high stiffness, lightweight, and complete recyclability are used in architectural panels and automotive components. Pure® is another 100% PP SRC which is available as tape, fabric, or sheet. Normal PP is heated and stretched to align the molecules resulting in a much stronger and stiffer material without any increase in density. It uses co-extruded tapes which exhibit a much larger processing window compared to Curve®. It has excellent impact properties and is used for ballistic and blast protection products. They are also suitable for automobiles, sports parts, and flat panels. Comfil® produces self-reinforced polyethylene terephthalate (PET) in the form of high-tenacity PET mixed with low melting PET, in the form of yarn, fabrics, consolidated plates, pellets for injection moulding and rods/tapes. They are also aiming to produce self-reinforced polyamides and self-reinforced polyolefins .

1.8 Applications

As mentioned earlier, SRPCs find a majority of the applications in automobiles, cargo containers, and to resist shock loads. Ultra-high-molecular-weight PE SRCs are used in biomaterial applications.

Tegris® (PURE®) and Curve® are the branded PP-based SRCs commercially available in the market for various applications shown in the figure. PP SRCs are widely accepted in automobile parts like underbody shields, interior panels, load floors, scuff plates, and panelling for trucks and vans. Lightweight of PP SRC helps to reduce weight and improve fuel efficiency with reduced exhaust emission.

Aerosplitters based on Tegris were introduced in 2011 and are used in NASCAR racing. They are also used in indoor panels replacing carbon fibre-reinforced composites. This is because of the lightweight and less cost with an advantage of reducing the injuries caused to the racer as it will not splinter when it breaks in accidents. Carbon fibre/Tegris/carbon fibre sandwich was also developed recently which has equal stiffness to a carbon fibre-only structure still lighter, more damage tolerant, and requires twice the energy to fail. Another is an aluminium/Tegris/aluminium sandwich construction, which takes three times the energy to fail.

Tegris is used as a protective armour by the U.S. military in its vehicles, primarily against Improvised explosive device (IEDs). There are also such diverse applications as small watercraft, helmets, outdoor furniture, and baggage. Impact resistance, stiffness, and weight are the primary concerns in hard shell luggage applications and PP SRCs are widely used by many manufacturers. Strengthening the fibre forms by nano-fillers is another option for future work. SRPC panels containing honeycomb or foam cores are being developed these days, and their properties can be improved by using hybrid SRCs, a combination of SRCs and non-SRCs. Fire resistance and heat conductivity issues are to be addressed with suitable remedies. Weldability of SRPC is another challenge and proper non-destructive testing to be developed for quality inspection.

There are not many processing methods available with SRPCs. Mathematical modelling of SRPC is still a major issue due to the difficulties in obtaining a proper relationship between the structure and its properties. Air cargo made of PE or PP SRCs would float on the ocean saving the luggage and lives in the event of an accident. Figure 1.2 represents some of the popular applications of SRPCs.

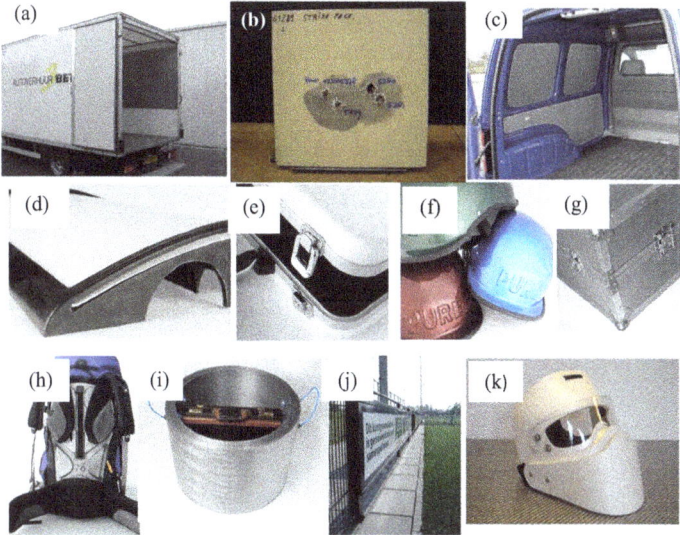

Figure 1.2: Some of the popular applications of self-reinforced composites.
(From top left to bottom right): (a) scuff plates of a container, (b) anti-ballistic panels, (c) impact panels of a mini-van, (d) promoflex – advertisement panels, (e) suitcase, (f) helmet, (g) flight cases, (h) sportsgear, (i) blast basket, (j) promodur – advertising panels, and (k) demining masks (source: www.ditweaving.com).

References

[1] L. M. Morgan, B. M. Weager, C. M. Hare and G. R. Bishop, "Self reinforced polymer composites: Coming of age," *17th Int. Conf. Compos. Mater.*, ID12.15, 2009.
[2] K. P. Matabola, A. R. De Vries, F. S. Moolman and A. S. Luyt, "Single polymer composites: A review," *J. Mater. Sci.*, 44, 23, 6213–6222, 2009.
[3] Á. Kmetty, T. Bárány and J. Karger-Kocsis, "Self-reinforced polymeric materials: A review," *Prog. Polym. Sci.*, 35, 10, 1288–1310, 2010.
[4] C. Gao, L. Yu, H. Liu and L. Chen, "Development of self-reinforced polymer composites," *Prog. Polym. Sci.*, 37, 6, 767–780, 2012.
[5] J. Karger-Kocsis and T. Bárány, "Single-polymer composites (SPCs): Status and future trends," *Compos. Sci. Technol.*, 92, 77–94, 2014.

[6]   J. Zou, G.-J. Zhang and Y.-M. Kan, "Formation of tough interlocking microstructure in ZrB2–SiC-based ultrahigh-temperature ceramics by pressureless sintering," *J. Mater. Res.*, 24, 7, 2428–2434, 2009.

[7]   Z. Luo, A. Lu, L. Han and J. Song, "In situ synthesis and properties of self-reinforced Si3N4–SiO2–Al2O3–Y2O3 (La2O3) glass–ceramic composites," *Bull. Mater. Sci.*, 40, 4, 683–690, 2017.

[8]   W. Zhang, M. Wang, W. Chen, Y. Feng and Y. Yu, "Evolution of inhomogeneous reinforced structure in TiBw/Ti-6AL-4V composite prepared by pre-sintering and canned β extrusion," *Mater. Des.*, 88, C, 471–477, 2015.

[9]   J. Li, Z. Yu, H. Wang and M. Li, "Microstructural evolution of titanium matrix composite coatings reinforced by in situ synthesized TiB and TiC by laser cladding," *Int. J. Miner. Metall. Mater.*, 17, 4, 481–488, 2010.

[10] S.-J. Park and S.-Y. Lee, "History and Structure of Carbon Fibers BT – Carbon Fibers," S.-J. Park, Ed. Dordrecht: Springer Netherlands, 1–30, 2015.

[11] G. R. Devi and K. R. Rao, "Carbon-carbon composites – an overview," *Def. Sci. J.*, 43, 4, 369–383, 1993.

[12] M. Bykowski, A. Hudgins, R. M. Deacon and A. R. Marder, "Failure analysis of the space shuttle Columbia RCC leading edge," *J. Fail. Anal. Prev.*, 6, 1, 39–45, 2006.

Chapter 2
Raw materials

Synthetic thermoplastic polymers like polyolefin (polyethylene (PE) and polypropyl-ene (PP)), acrylics, polyesters, and polyamides are the usual source for self-reinforced polymer composites (SRPC). The details of commercially available raw materials and their properties are elucidated in the beginning of this chapter along with various pos-sible combinations [1–4]. There is a lot of scope for the fabrication of raw materials with desired properties. Thus, some of the existing manufacturing procedures of raw materials are also included in this chapter.

2.1 Polyethylene

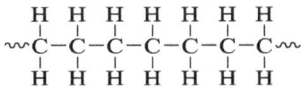

Figure 2.1: Polyethylene or polythene.

Ethylene monomer ($CH_2 = CH_2$, IUPAC name: ethene) is polymerized to form PE. Eth-ylene is a two-carbon compound with two hydrogen atoms on each carbon. Although the classification chart of PE (Figure 2.1) indicates only three types of PE forms, based on their density which are again subdivided into plenty of different combinations.

Ethylene is a colourless and flammable unsaturated hydrocarbon. The stable double bond of the ethylene is broken to form long chains of PE in the presence of a catalyst. Various grades of PE which differ in molecular weight (ranging from 28,000 to 280,000) and branching can be obtained by proper conditions of temperature, pressure, or catalysis. Three different forms of PE are the most common grades manu-factured in bulk: low-density PE (LDPE, <0.930 g/cm³), linear low-density PE (LLDPE, 0.915–0.940 g/cm³), and high-density PE (HDPE, 0.940–0.965 g/cm³).

PE being a widely accepted popular polymer material, it is used in many consumer products and industrial applications like plastic bottles, plastic bags, film packaging, electrical insulation, extruded plastic piping, food packaging, and cable coating.

2.1.1 Types of polyethylene

Different grades of PE are given as follows:
(1) Ultra-high-molecular-weight PE (UHMWPE)
(2) Ultra-low-molecular-weight PE (ULMWPE)
(3) HDPE

https://doi.org/10.1515/9783110647334-002

(4) Cross-linked PE (PEX or XLPE)
(5) Medium-density PE (MDPE)
(6) LLDPE
(7) LDPE
(8) Very-low-density PE (VLDPE)
(9) Chemically modified PE

2.1.1.1 Ultra-high-molecular-weight polyethylene (UHMWPE)

The molecular weight of UHMWPE is the highest among other types with a molecular mass ranging from 3.5 to 7.5 million amu. Though UHMWPE can be prepared through any of the catalyst technologies, Ziegler catalyst is the most common. UHMWPE has a wide range of applications because of its exceptional toughness, cut, wear, and chemical resistance. Consumer products like cans, various bottle handling machine parts, gears, and bearings are some of its applications. Auricular parts of implants that are used for hip and knee replacements are constructed using UHMWPE. UHMWPE fibres are also used in bulletproof vests like aramid fibres because of their excellent ballistic impact properties (Figure 2.2).

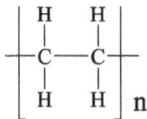

Figure 2.2: Monomer structure of UHMWPE.

2.1.1.2 Ultra-low-molecular-weight polyethylene (ULMWPE)

This grade of PE is also known as PE wax. It exists in the solid or semi-solid form at room temperature. Melting point is around 150 °C. Emulsifiable and non-emulsifiable PE wax are available with molecular weight ranging from 2,000 to 4,000 and density ranging from 0.92 to 0.96.

2.1.1.3 High-density polyethylene (HDPE)

HDPE (density >0.940 g/cm^3) is used widely due to its appreciable mechanical properties. Stronger intermolecular forces exist in linear chains than branched chains due to the close packing of molecules. HDPE is made by chromium/silica catalysts, Ziegler–Natta catalysts, or metallocene catalysts. Detailed synthesis is discussed in the next chapter. It is used in products and packaging such as utility vessels, garbage containers, toys, and water pipes (Figure 2.3).

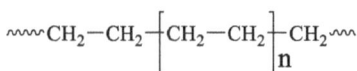

Figure 2.3: Structure of HDPE.

2.1.1.4 Cross-linked PE

PEX contains bonds with cross-links, and their density varies from a medium to high. These cross-linking transforms the thermoplastic into a thermoset, thereby improving the high-temperature property of these polymers. This is done by reducing its flow and enhancing the chemical resistance.

2.1.1.5 Medium-density PE

MDPE has a density range of 0.926–0.940 g/cm^3 and is produced by chromium/silica catalysts, metallocene catalysts, or Ziegler–Natta catalysts. MDPE has good shock and drop resistance properties. It has better stress-cracking resistance and less notch-sensitivity than HDPE. MDPE is used in gas pipes and fittings, shrink film, packaging films, sacks, screw closures, and carrier bags.

2.1.1.6 Low-density polyethylene (LDPE)

High pressure during manufacturing results in LDPE (density 0.910–0.940 g/cm^3) to have a high degree of short- and long-chain branching; therefore, the chains do not pack into the crystal structure. The instantaneous dipole and induced dipole attraction are less causing less strong molecular forces. This results in a lower tensile strength and increased ductility. LDPE is produced by free-radical polymerization. The molten LDPE gets unique and desirable flow properties due to long chains and high degree of branching (Figure 2.4).

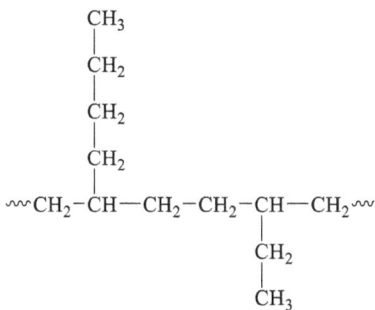

$$
\begin{array}{c}
CH_3 \\
| \\
CH_2 \\
| \\
CH_2 \\
| \\
CH_2 \\
| \\
\text{---}CH_2\text{--}CH\text{---}CH_2\text{--}CH_2\text{--}CH\text{---}CH_2\text{---} \\
\hspace{4cm} | \\
\hspace{4cm} CH_2 \\
\hspace{4cm} | \\
\hspace{4cm} CH_3
\end{array}
$$

Figure 2.4: Structure of LDPE.

2.1.1.7 Linear low-density polyethylene (LLDPE)

LLDPE has a density range of 0.915–0.925 g/cm^3. LLDPE is a substantially linear polymer with significant numbers of short branches, commonly made by copolymerization of ethylene with short-chain alpha-olefins (e.g. 1-butene, 1-hexene, and 1-octene). LLDPE exhibits higher impact and puncture resistance and tensile strength than LDPE. Though it is a difficult process, lower thickness (gauge) films can be blown, with better environmental stress-cracking resistance. LLDPE is used in packaging, cable coverings, containers, toys, buckets, lids, and pipes. While another application exists, LLDPE is used predominantly in film applications due to its flexibility, toughness, and relative transparency. Product examples range from agricultural films, bubble wrap, to multilayer and composite films (Figure 2.5).

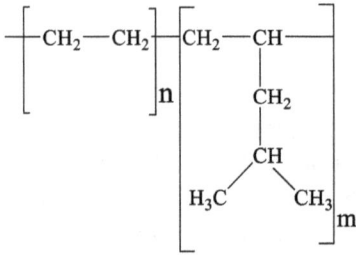

Figure 2.5: Structure of LLDPE.

2.1.1.8 Very-low-density polyethylene (VLDPE)

With a density range of 0.880–0.915 g/cm^3, VLDPE is substantially a linear polymer with high levels of short-chain branches, commonly manufactured by copolymerization of ethylene with short-chain alpha-olefins (e.g. 1-butene, 1-hexene, and 1-octene). Due to the greater co-monomer incorporation exhibited by metallocene catalysts, VLDPE is most commonly produced using those catalysts. VLDPEs are used for hose and tubing, ice and frozen food bags, food packaging, and stretch wrap as well as impact modifiers when blended with other polymers.

2.1.1.9 Copolymers

Ethylene copolymerized with a number of monomers like vinyl acetate and acrylates, are used in packaging and sporting goods, and a variety of applications. PE modified in the polymerization by polar or non-polar co-monomers or by cross-linking, chlorination, and sulfochlorination are also available in the market.

2.1.1.10 Cross-linked PE

Some of the properties of PE like resistance against environmental stress crack resistance can be improved by cross-linking PE by various methods. Properties like shear modulus, abrasion resistance, and impact strength can also be improved, whereas hardness and rigidity are considerably reduced. PE-X is thermally resistant even up to 250 °C without mechanical load. PE-X is used as an insulating material for hot water pipes, moulded parts in electrical engineering, medium, and high voltage cable insulation, plant engineering, and in the automotive industry. There are various methods for cross-linking depending upon the requirement of desirable property. PE-X piping systems are the most robust system available in the market (Figure 2.6).

The degree of cross-linking has very much importance in the properties of cross-linked material. A lower rate of cross-linking leads to shifting the material

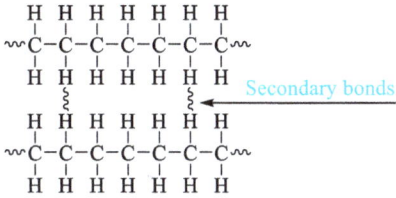

Figure 2.6: Structure of cross-linked polymer.

into the behaviour of elastomers and a higher rate brings thermosetting behaviour into the material. There are four types of cross-linked PE which are denoted as PE-Xa, PE-Xb, PE-Xc, and PE-Xd.

2.1.1.10.1 Peroxide cross-linking (PE-Xa)

Highly flexible and softer peroxide cross-linked PE can be prepared by peroxide cross-linking of PE. Dicumyl peroxide, di-tert-butyl peroxide, and di-tert-amyl peroxide are some of the examples of peroxides used in the cross-linking process. Engel process is used for the cross-linking of PE with peroxides. In this process, HDPE and 2% peroxide are mixed at low temperatures in an extruder and then the temperature is raised between 200 and 250 °C. At this high temperature, the peroxide decomposes to peroxide radicals, which removes hydrogen atoms from the polymer chain. These radicals combine to form a cross-linked structure. About 75% is the degree of cross-linking to be achieved.

2.1.1.10.2 Silane cross-linking (PE-Xb)

Silanes are a typical class of inorganic with the chemical formula Si_nH_{2n+2}. There are two types of processes for silane cross-linking. The degree of cross-linking required is 65% in this case. One is a single-step process and another is a double-step process. These processes are also called silane grafting. In a single-step process, a copolymer of silane and PE may be formed, or free-radical generation and cross-linking take place in a single stage. In the double-step process, vinyltrymethoxysilane is grafted on a PE chain after treatment of PE with peroxide.

2.1.1.10.3 Radiation cross-linking (PE-Xc)

In this type of cross-linking, PE is exposed to controlled gamma or beta radiation by using an electron accelerator or isotopic radiator. PE absorbs radiation energy, and chemical bonds are broken to form free radicals. These radicals react with new chemical bonds to form a network. Most of the polymers that can be cross-linked chemically can also be cross-linked by using this process. Partial cross-linking is also possible by covering some of the parts of the material. Low-temperature operation is the main advantage of this type of cross-linking. The favourable degree of cross-linking is 60% for radiation cross-linking.

2.1.1.10.4 Azo cross-linking (PE-Xd)

This series was prepared by using Lubonyl process and not existing much in the industry.

2.1.1.11 Chlorination and sulfochlorination

These are the two important chemical modifications of PE. Chlorinated PE is an inexpensive elastomer material having chlorine content from 25% to 42%. Chlorinated PE elastomers and resins have excellent physical and mechanical properties and compression set resistance, flame retardancy, tensile strength, and abrasion resistance. It is used in automotive and industrial hose and tubing, moulding, and extrusion. Application is based on the level of degree of cooperated chlorine during the chlorination process.

Chlorosulfonated PE is obtained by the chemical modification of PE with chlorine and sulfur dioxide. It has a density of 1.11–1.26 g/cm^3, chlorine content of 25–47%, and sulfur content of 0.8–2.2%, and is used as the preliminary material for ozone-resistant synthetic rubber. It is resistant to fire, to oil, and to the action of microorganisms and exhibits good adhesion to various surfaces. It is insoluble in aliphatic hydrocarbons and alcohols, slightly soluble in ketones and esters, and readily soluble in aromatic hydrocarbons, such as toluene and xylene, and chlorinated hydrocarbons. Chlorosulfonated PE is superior to other rubbers in its resistance to the effects of ozone and inorganic acids, such as chromic, nitric, sulfuric, and phosphoric acids, as well as to the effects of concentrated alkalies, chlorine dioxide, and hydrogen peroxide. It is resistant to light, is impermeable to gas, and has good dielectric properties.

2.1.1.12 Biobased polyethylene

PE can be produced from ethanol which is synthesized by fermentation of biodegradable renewable natural resources like sugar cane and wheat grain. By chemical dehydration reaction, ethanol is converted to ethylene, and PE can be produced by the conventional polymerization reaction. PE produced by this method is also called renewable PE. Resulting PE is chemically similar to the conventional PE produced from crude oil resources and possesses identical properties. It is non-biodegradable and renewable in the same methods adopted for recycling conventional PE grades.

2.1.2 Synthesis of PE

PE was first synthesized by accident in 1898 while heating diazomethane. Later, free-radical process was discovered in 1930s which helped in the industrial production of PE. Pressure plays an important role in the formation of various grades of PE. High-pressure synthesis is required for the production of LDPE, while low-pressure operations are used for the production of HDPE.

HDPE and LLDPE are prepared at low temperature and pressure by Ziegler–Natta polymerization method* proposed by Karl Ziegler of Germany and Guilio Natta of Italy. The catalyst used in this process is popularly known as Ziegler–Natta catalyst. Ziegler–Natta catalyst is a combination of a transition metal compound of an element from groups IV to VIII, and an organometallic compound of a metal from groups I to III of the periodic table. The transition metal compound is referred to as the catalyst and the organometallic compound as the co-catalyst. The most common catalysts consist of titanium compounds with aluminium alkyl. Based on solubility, these can be classified into two categories [1–4]:

(i) Heterogeneous catalysts: These are based on titanium or vanadium compounds used for polymerization reactions, usually in combination with organo-aluminium compounds like tri-ethyl aluminium ($Al(C_2H_5)_3$) as co-catalysts. These are widely used in industries.

(ii) Homogeneous catalysts: These are another broad class of catalysts and are based on complexes of Ti, Zr, or Hf. They are generally used in combination with a range of different organo-aluminium co-catalysts known as metallocene/methyl aluminoxane (MAO).

LDPE is prepared at high pressure with free-radical initiators. These high-pressure operations lead to the production of branched PE due to intermolecular and intra-molecular chain transfer during polymerization.

2.1.2.1 Synthesis of LDPE

LDPE (<0.930 g/cm^3) has a high degree of short- and long-chain branching. It is manufactured at high pressure (1,000–3,000 atm) and at moderate temperature (420–570 K) by free-radical polymerization.

(i) Free-radical polymerization
A free radical is a molecule with at least one unpaired electron. Free-radical poly-merization is done at very high pressure and at an elevated temperature. It is a type of chain-growth polymerization. Initiation, propagation, and termination are the three different steps involved in the entire process (Figure 2.7).

Presence of trace of oxygen is required as the initiator. Oxygen reacts with some of the ethane to form highly reactive organic peroxide molecules due to the presence of weak oxygen–oxygen single bonds and breaks easily to produce free radicals. The remaining free radicals also react with the ethylene molecules produc-ing long-chain free radicals resulting in the propagation of polymer chains. When

---

* For the polymerization of ethylene into high-molecular-weight HDPE at room temperature, Karl Ziegler discovered the catalyst based on titanium tetrachloride ($TiCl_4$) and diethylaluminium chloride [$(C_2H_5)_2AlCl$] as a co-catalyst in 1953. Giulio Natta used this polymer to polymerize propylene into crystalline PP. Karl Ziegler and Giulio Natta received the Nobel Prize in 1963 for this contribution.

two free radicals react each other, the chain terminates as there is no free radical left to react further.

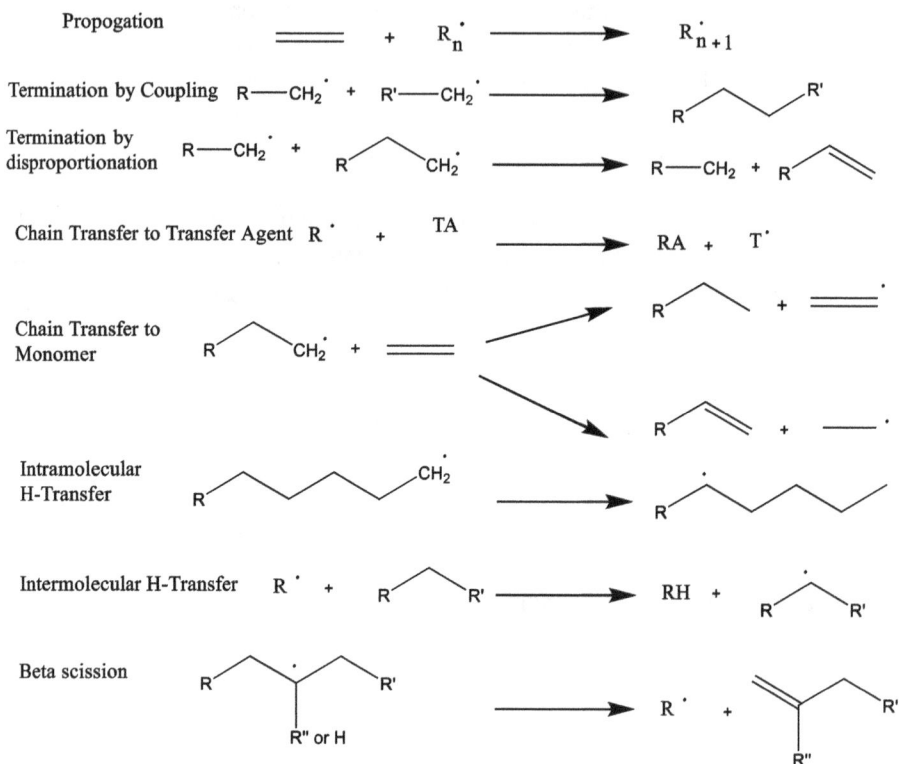

Figure 2.7: Free-radical polymerization.

2.1.2.2 Synthesis of HDPE

There are two types of catalysts used in the manufacturing process of HDPE. One of them is Ziegler–Natta catalyst which was explained earlier.

Another common catalyst is an inorganic catalyst called Phillips catalyst, prepared by depositing chromium(VI) oxide on silica.

The manufacturing process of HDPE can be further classified based on the process.

(i) Slurry polymerization or suspension polymerization/solution polymerization (Ziegler–Natta polymerization)

Initially, titanium tetrachloride reacted with an metal alkyl in the presence of an inert solvent at a temperature between 100 and 130 °C and at a pressure between 1 and 20 atm. Ethylene–hydrogen mixture is introduced into the reaction vessel in the gaseous state. Ethylene reacts with the catalyst in a long loop reactor under continuous stirring to form PE. Presence of solvent helps in dissipating the heat. As the melting point of PE is 130 °C, PE formed at this state is in solid phase. Then the catalyst is deactivated in the presence of an alcohol. Through a filtration and drying process, PE is recovered (Figure 2.8).

In slurry polymerization, the catalyst granules are mixed with liquid hydrocarbon like isobutene or hexane. In solution polymerization, ethene and hydrogen are passed into the catalyst solution in a hydrocarbon under pressure. Remaining processes are similar in both the methods.

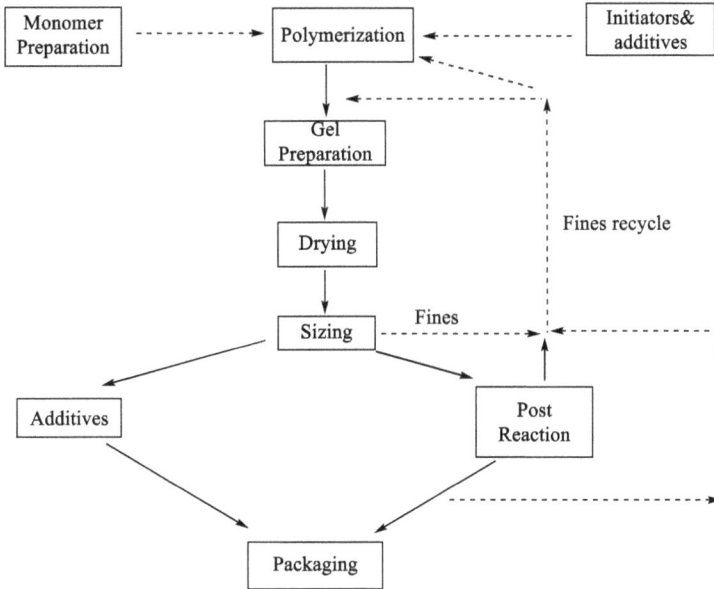

Figure 2.8: Solution polymerization.

(i) Gas-phase polymerization (Philips' catalyst)
The mechanism of these types of polymerization is similar to that of the Ziegler–Natta polymerization. The reactive mechanism is also known as anionic polymerization or living polymerization. The pressure is higher than the Ziegler–Natta polymerization (30 atm). Increased pressure causes less branching in the HDPE and thus high density compared to that produced by Ziegler–Natta catalyst (Figure 2.9).

Chromium oxide on high surface area silica is used as a very active catalyst in this process. Ethylene is passed in a loop reactor along with α-olefin. The catalyst and the inert solvent are introduced into the reactor. An active site of chromium carbon bond is formed when the catalyst reacts with α-olefin. As the reaction is highly exothermic, inert solvent and a cooling jacket take the function of dissipating the heat generated. The active sites on the catalyst induce the chain development both outward and inward. The polymer is dried and pelletized after removing the solvent by hot flashing. The conversion percentage is very high in this method (95–98%); thus, the recovery process of unreacted ethylene is eliminated.

PE formed by this process is more crystalline and used for making durable products. Molecular weight of HDPE can be controlled by pressure, temperature, chain transfer reagents, or catalytic control.

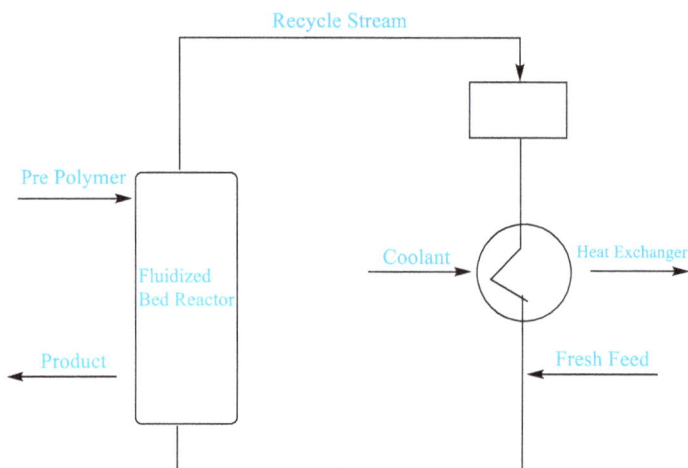

Figure 2.9: Gas-phase polymerization.

2.1.2.3 Synthesis of LLDPE

In SRPCs, density plays an important role as it affects the melting temperature. Thus, LDPE has an important role in self-reinforced polymerization. But the high-pressure requirement for the manufacturing increases the capital cost. LLDPE is developed by using both the Ziegler–Natta catalyst and the inorganic catalysts, and any of the methods described above for the production of HDPE can be used for the manufacture of LLDPE based on which catalyst is introduced, as gas-phase polymerization process is used when inorganic catalyst is used. Additionally, but-1-ene or hex-1-ene is introduced to the reactor, and these monomers carry a few branches which reduce the density.

2.2 Polypropylene

PP is a thermoplastic "addition" polymer with chemical formula $(C_3H_6)_n$ synthesized from propylene monomers (Figure 2.10). It was first synthesized in 1951 by Paul Hogan and Robert Banks. Commercial production of PE was triggered in 1954 by Nobel laureate Giulio Natta in Spain by using the method proposed by Ziegler for the production of PE. PP is also one of the important polymers used in daily life with a wide variety of applications in automotive components, packaging and labelling, ropes, carpets, clothes, stationery, plastic parts and reusable containers of various types, laboratory equipment, loudspeakers, and polymer banknotes. PP is the second most important plastic with revenue exceeding US \$145 billion in 2019. The sales of this material are expected to grow at a rate of 5.8% per year until 2021 [5–8].

$$\left[-CH_2\text{-}\underset{\underset{CH_3}{|}}{CH}-\right]_n$$

Polypropylene   **Figure 2.10:** Structure of PP.

2.2.1 Types

Homopolymers and copolymers of PP are available in different grades dependent on the application. PP is also classified into three types based on tacticity (relative position of methyl group on the parent carbon chain): isotactic (all methyl groups are on the same side), syndiotactic (methyl groups are on the alternate sides of the chain on the nearby carbon atom), and atactic (methyl groups are randomly distributed on the parent carbon chain) (Figure 2.11).

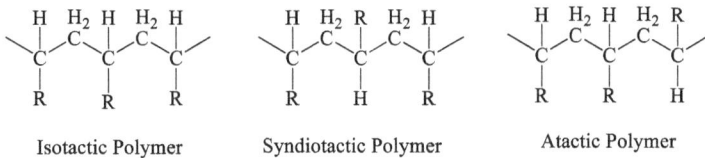

Isotactic Polymer   Syndiotactic Polymer   Atactic Polymer

Figure 2.11: Types of PP based on stereochemistry.

2.2.2 Other types of PP

Adding ethylene–propylene rubber to PP homopolymer increases its low-temperature impact strength. When randomly polymerized, the ethylene monomer added to PP homopolymer will decrease the polymer crystallinity, lower the melting point, and make the polymer more transparent.

2.2.2.1 PP maleic anhydride
The functionalization of PP by polar monomers in order to increase the polarity of PP and to improve its affinity with other polar materials is the advantage of PP maleic anhydride. Maleic anhydride is used for this purpose, owing to the higher reactivity of the anhydride group to successive reactions.

2.2.2.2 High crystalline polypropylene

High crystalline PP resins are having higher stiffness and an extremely high isotactic index, excellent chemical, heat resistance, and good processability. High crystalline PP can be converted with existing extrusion or injection moulding equipment.

2.2.2.3 Random copolymers

Random copolymers are obtained by the modification of the PP chain by adding small amounts of other monomers in order to achieve improved hot sealing characteristics and impact resistance when compared with homopolymer. Random copolymer PP resins exhibit high chemical resistance against most inorganic acids, alkalis and salts, excellent optical properties, and environmental stress cracking resistance. These resins have wide applications as skin layer in extrusion for flexible packaging and in injection-moulded consumer products.

2.2.2.4 Block copolymers (HECO)

In order to achieve good stiffness and low-temperature impact performance, PP matrix is polymerized with a rubber component resulting in block copolymers. They are used in microwave ovens, automotive bumpers, pails and crates, and sewage pipe systems due to high-temperature applications and high impact strength. As there is a high flexibility in available viscosity range of these materials, block copolymers can be processed through injection moulding, extrusion methods, or any other processes.

2.2.2.5 Biaxially oriented polypropylene (BOPP)

In order to improve strength and lucidity of PP unlike other grades, the type of PP obtained by extruding and stretching of PP film is called biaxially oriented PP (BOPP). They are used in packaging material for artistic and retail products. It is easy to coat, print, and laminate and customize its appearance and properties.

2.2.2.6 Expanded polypropylene (EPP)

Expanded polypropylene (EPP) has very good impact characteristics due to its low stiffness; this allows EPP to resume its shape after impacts. EPP is extensively used in model aircraft and other radio controlled vehicles by hobbyists.

Properties of PP are comparable with that of PE. Molecular weight, crystallinity, tacticity, and others are the important factors affecting the properties of PP. Crystallinity of PP is in between that of LDPE and high-density polyethylene. Copolymerization with polyethylene is done in order to achieve better mechanical properties and chemical resistance. Melt flow index (rate at which the melt flow during processing) is dependent on molecular weight. They have high melt flow rate which helps in moulding though it adversely affects other physical properties like impact strength.

PP is an economical commodity plastic with the lowest density (0.895 and 0.92 g/cm) with which lower weight and certain mass of plastic can be manufactured. The Young's modulus of PP is between 1,300 and 1,800 N/mm². PP is normally tough and flexible and has good resistance to fatigue. PP is less chemically reactive compared to PE and resists almost all organic solvents, apart from strong oxidants at room temperature. PP can be dissolved in low polarity solvents (e.g. xylene, tetralin, and decalin) at an elevated temperature. Crystallinity of isotactic PP is in between that of LDPE and HDPE. Isotactic and atactic PP is soluble in p-xylene at 140 °C. Isotactic PP precipitates when the solution is cooled to 25 °C but the atactic PP portion remains soluble in p-xylene.

2.2.3 Synthesis of PP

Propylene or propene (IUPAC) or methyl ethylene (C_3H_6, $CH_2 = CH_2$-CH_3) is the monomer of PP. Propene is produced from gas oil, propane, naphtha, and ethane. Double bond in the propene molecule breaks in the polymerization process forming long chains of PP. In the production process, the monomer is subjected to heat and pressure in the presence of catalyst. Controlling these operating conditions will yield different grades. The processes can be slurry process, solution process, or gas-phase process. As every propene molecule carries a methyl group (-CH_3), orientation of this branch in the polymer molecule has a great role in deciding the crystallinity of the polymer. Relative orientation of methyl groups in the polymer chain which affects the properties of the polymer is referred to as tacticity.

PP can be isotactic, syndiotactic, or atactic. In isotactic PP, all the methyl groups are placed on the same side of the carbon chain. Methyl groups on alternate carbon atoms orient in opposite directions to form syndiotactic PE. Atactic PP is formed by the random orientation of methyl groups. This difference is caused based on which face of the propylene comes in co-ordination with the catalyst in synthesis.

Synthesis of these three types of PP is also different. Isotactic PP has more crystallinity, which causes more stiffness and higher melting temperature (171 °C). Ziegler–Natta polymerization is capable of yielding unidirectional-oriented PP chains. A combination of Ziegler–Natta catalyst and metallocene catalysts is also used in commercial production of isotactic PP. Syndiotactic PP is produced by another type of metallocene catalysts. Atactic PP may be made either by Ziegler–Natta catalysts or metallocene catalysts. After the discovery of Ziegler–Natta catalyst, Giulio Natta produced PP with the catalyst suggested by Ziegler. Commercial production of PP commenced in 1957 in Italy, Germany, and the USA. Natta and Breslow discovered the metallocene catalyst to catalyse olefin polymerization with conventional co-catalyst (Al alkyls). Later, PP became a material for manufacturing various products like fibres, fabrics, and non-woven fabrics on a commercial scale. Second-generation Ziegler–Natta catalysts was introduced in 1973 with $TiCl_3$ purple phases at lower temperatures. After 1975, many companies commercialized third-generation catalysts supported on $MgCl_2$. Kaminsky and Sinn discovered high-activity metallocene single-site catalysts (SSCs) using MAO as co-catalyst during 1977–1980.

In 1984, Ewen at the Exxon Company (USA) demonstrated that appropriate titanocenes render partially isotactic PP. Fourth-generation Ziegler–Natta catalysts based on aluminium-oxane activated metallocene complexes used in 1991. Later, Brookhart and co-workers discovered non-metallocene SSC based primarily on chelated late transition metals.

Lyondell Basell commercialized PP based on fifth-generation Ziegler–Natta catalyst that used 1.3-diethers and succinate as donors in 1997.

Majority of PP polymers used in the industry are either isotactic or syndiotactic PP due to the desirable properties. There were several stages in the development of synthesis of PP. The process is to bring monomers into contact with catalysts which have active sites for polymerization. PP formed due to the polymerization process at the active sites precipitates out. Residual catalyst, solvent, and atactic PP formed are to be removed. The process of removal of catalyst is known as deashing. Earlier method of polymerization was solvent polymerization or slurry polymerization in which PP particles were dispersed in the form of slurry in the solvent. Deashing, solvent, and atactic polymer removal are eliminated in the next generations of development in the process. Gas-phase polymerization is the most economical and widely used modern technology in the industrial production of PP [5–8].

2.2.3.1 Ziegler–Natta polymerization

Catalyst used in this process is titanium(IV) chloride and aluminium alkyl (e.g. triethyl aluminium). The mostly followed commercial process for the production of isotactic PP in Ziegler–Natta polymerization is gas-phase polymerization and bulk polymerization even though slurry polymerization is an alternate process.

(i) Bulk process or liquid-phase polymerization process
Liquid propene is treated with the catalyst at a pressure of 30–40 atm and a temperature of 50–80 °C which keeps the liquid phase of polypropene in the entire process and additional solvents are not required. This reduces the cost for the recovery of solvents. PP can be produced in the powder form. Unreacted monomer can be recycled (Figure 2.12).

(ii) Gas-phase or vapour-phase polymerization process
In this process, polymerization is carried out in the gas phase in a temperature range of 40–85 °C at a pressure of 8–35 atm. The process for removing atactic PP is eliminated due to the high performance of catalysts. The catalyst remaining in the process is eliminated by using alcohol or water (Figure 2.13).

All developments in the synthesis of PP is aiming at reducing the heterogeneousness in the polymer particle. This can be obtained by using a circular-type reactor or by providing a reactor with a narrow retention time distribution.

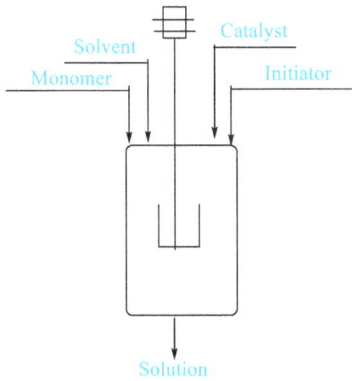

Figure 2.12: Liquid-phase polymerization.

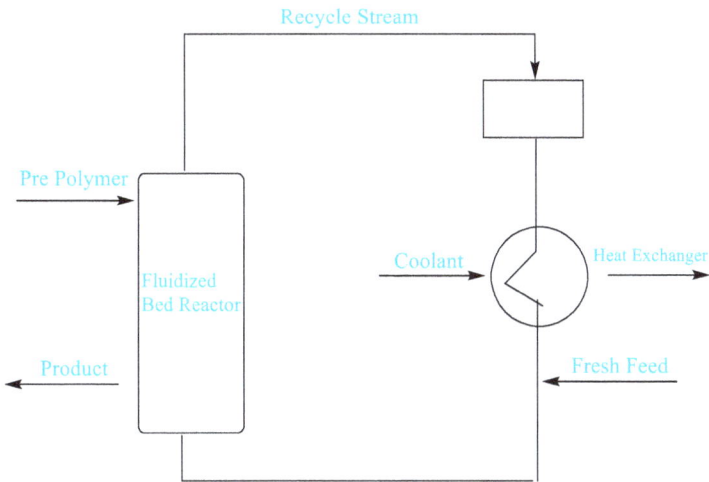

Figure 2.13: Gas-phase polymerization.

2.2.3.2 Metallocene catalytic polymerization

Even though Ziegler–Natta catalyst is widely used for the production of isotactic PP, most of the commercially available isotactic PPs manufactured by this process may be a mixture of short atactic chains and other chains too. It is challenging to produce isotactic PP with varying molecular weight by keeping constant isotacticity.

The drawback when using Ziegler–Natta isotactic PPs lies in the fact that some structural parameters are almost impossible to study separately; for example, molecular weight and tacticity are strongly coupled in these polymers. In fact, isotactic Ziegler–Natta PPs can be considered as a mixture of very different types of chains: short atactic chains are present even if it is impossible to evaluate the separated effect of molecular weight and tacticity and tacticity distribution. However, because of the presence of only one active centre in metallocene catalysts, metallocene PPs are more and the chains resemble one another much more than when using Ziegler–Natta catalysts. PE and PP produced by metallocene catalysts also possess uniform microstructures with narrow molecular weight distribution and chemical composition distribution.

2.3 Nylon

Nylon is a thermoplastic condensation polymer and aliphatic or semi-aromatic polyamide made of repeating units linked by peptide bonds. They can be melt-processed into fibres, films, or shapes. Nylon is one of the largest engineering polymer families; the global demand of nylon market is expected to reach US $30 billion in 2020 with an average annual growth of 5.5%.

Nylon as well as its copolymers are commercially synthesized by reacting difunctional monomers containing equal parts of amine and carboxylic acid, to form amides at both ends of each monomer. A wide variety of additives can also be mixed to obtain many different property variations in nylon polymers (Figure 2.14).

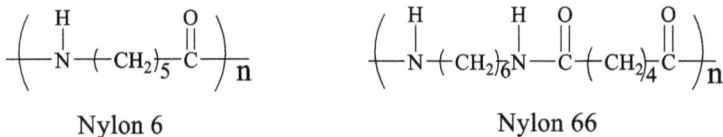

Nylon 6   Nylon 66

Figure 2.14: Types of nylon.

2.3.1 Types of nylon

2.3.1.1 Homopolymers
Nylon 6 (PA 6) $[NH-(CH_2)_5-CO]_n$ made from ε-caprolactam processes at a lower temperature (maximum of 220 °C) possesses a minor mould shrinkage. Nylon is lightweight and has good toughness, impact resistance, and surface finish. It is used in various applications like automotive components, firearm components, circuit breakers, pulleys, and gears [10–14].

Nylon 66 (PA 66) $[NH-(CH_2)_6-NH-CO-(CH_2)_4-CO]_n$ made from hexamethylenediamine and adipic acid has a higher melting point (around 265 °C), making it suitable for higher temperature applications. Nylon 66 has improved stiffness, higher tensile, flexural modulus, high melting point, and durability. Nylon 66 has wide applications in battery modules, bolts and fasteners, general-purpose housings, and so on.

Examples of commercially available homopolymers are PA46 DSM stannyl, PA410 DSM Ecopaxx, PA4T DSM Four Tii, and PA66 DuPont Zytel.

2.3.1.2 Copolymers

Copolymers of PA have lower crystallinity and therefore lower the melting point. One of the copolymers nylon 6/66 (PA 6/66) $[NH\text{-}(CH_2)_6\text{-}NH\text{-}CO\text{-}(CH_2)_4\text{-}CO]_n\text{-}[NH\text{-}(CH_2)_5\text{-}CO]_m$ is made from caprolactam, hexamethylenediamine, and adipic acid, while nylon 66/610 (PA66/610) $[NH\text{-}(CH_2)_6\text{-}NH\text{-}CO\text{-}(CH_2)_4\text{-}CO]_n\text{-}[NH\text{-}(CH_2)_6\text{-}NH\text{-}CO\text{-}(CH_2)_8\text{-}CO]_m$ is made from hexamethylenediamine, adipic acid, and sebacic acid. PA6/66 DuPont Zytel, PA6/6 T BASF Ultramid T, PA6I/6 T DuPont Selar PA, PA66/6 T DuPont Zytel HTN, and PA12/MACMI EMS Grilamid TR are examples of the commercially available copolymers. Nylon polymers are miscible with its other grades to form variety of blends. Transamidation is another process by which random copolymers are formed by the reaction of amine and amide groups. Polyamides have wide variety of crystallinity including amorphous forms. PA46 and PA66 are highly crystalline, PAMXD6 is low crystalline made from m-xylylenediamine and adipic acid, PA6I made from hexamethylenediamine and isophthalic acid is amorphous, and PA66 is an aliphatic semi-crystalline homopolyamide.

Most common types of nylon are nylon 6, and nylon 6,6. Nylon 66 has more compact molecular structure, better weathering properties, better sunlight resistance,, and excellent abrasion resistance. While nylon 6 is easy to dye, it more readily fades, and exhibits a higher impact resistance, a more rapid moisture absorption, greater elasticity, and elastic recovery. Its high tenacity fibres are used for seatbelts, tire cords, ballistic cloth, and so on. Nylon 6 also possesses excellent abrasion resistance, high resilience, and resistance to chemicals, insects, fungi, and animals. Nylons are hygroscopic, and absorb or desorb moisture as a function of the ambient humidity, still less absorbent than wool or cotton. Moisture content has a significant role in properties of polymers like electrical properties. Dry nylon is a good electrical insulator.

2.3.2 Synthesis of nylon

Polyamides are polymers containing repeating amide (-CONH) groups. There are natural and synthetic polyamides. Protein is one of the natural polyamide. Synthetic polyamides can be aliphatic or aromatic polyamides. Most popular aliphatic synthetic polyamides are nylon and aromatic polyamides are Kevlar and carbamide plastics.

Nylon is the widely used polyamide in industries and daily life. There are many varieties of nylon available in the market. But nylon 6 and nylon 6,6 contribute to 90% of the global market share. Thus, we elaborate the synthesis of these two grades with more importance in this session. Both these polyamides can be manufactured by converting benzene to cyclohexane [9–12].

2.3.2.1 Nylon 6

When benzene is treated with hydrogen in the presence of nickel catalyst under pressure, cyclohexane is formed. Cyclohexane is oxidized to form cyclohexanone and cyclohexanol (in KA mixed oil, K stands for ketone and A stands for alcohol). For nylon 6, pure cyclohexanone is required. The mixture of ketone and alcohol is heated under pressure with copper(II) and chromium(III) oxides and dehydrogenated to cyclohexanone. Cyclohexanone is then converted to caprolactam by the reaction with hydroxylamine hydrogen sulfate. This caprolactam, water (catalyst), and a molecular mass regulator are heated under nitrogen at 500 K for about 12 h to form the polymer (Figure 2.15).

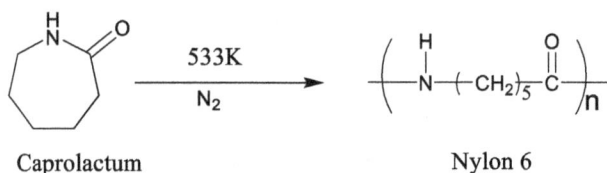

Caprolactum   Nylon 6

Figure 2.15: Synthesis of nylon 6.
*Asahi process is the recent method for producing cyclohexanol from benzene by hydrogenation and subsequent hydration.

2.3.2.2 Nylon 6,6

Nylon 6,6 is produced by condensation polymerization of 1,6-diaminohexane (hexamethylenediamine) with hexanedioic acid (adipic acid). Hexanedioic acid can be obtained from the first steps in the process of production of nylon 6,6. KA mixed oil (refer) is oxidized to form in the liquid phase with 60% concentrated nitric acid at 330 K in the presence of catalysts. Ammonium vandate(V) and copper(II) nitrate are the catalysts used. Hexamethylenediamine is produced from buta-1,3-diene and propenonitrile (polyacrylonitrile).

The acid and the diamine are then heated together to form a salt.

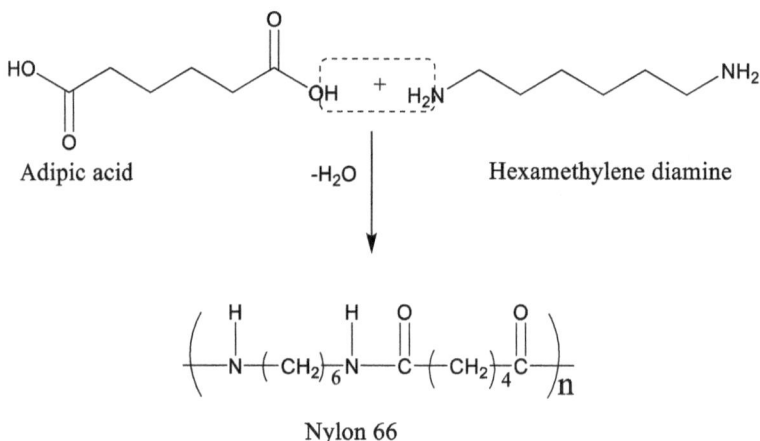

Adipic acid $-H_2O$ Hexamethylene diamine

Nylon 66

Figure 2.16: Synthesis of nylon 66.

The process of obtaining polymer is explained below. The chain length is regulated by controlling process conditions such as reaction time, temperature, and pressure (Figure 2.16).

2.4 Polyester

Polyester is a polymer with the ester functional group in the polymer chain formed of the monomer ethylene terephthalate, with ($C_{10}H_8O_4$) as repeating units. Polyester can be thermoplastic or thermosets depending on their chemical structure. Most common and widely acceptable type of polyester for processing into self-reinforced polymer in this family is polyethylene terephthalate (PET) [15–16]. PET is the most common semi-crystalline thermoplastic polyester and widely used in fibres for clothing, bottles, seat belts, and containers. It is the fourth most produced polymer after PE,, PP, and polyvinyl chloride (PVC).

More than 60% of worldwide production of PET is for synthetic fibres and 30% is for production of plastic bottles. Depending upon their crystal structure and particle size, transparent or opaque varieties are also available.

PET may exist both as an amorphous (transparent) and as a semi-crystalline polymer. Its monomer bis(2-hydroxyethyl) terephthalate is synthesized by the esterification reaction between terephthalic acid (TPA) and ethylene glycol with water as a by-product, or by transesterification reaction between ethylene glycol and dimethyl terephthalate with methanol as a by-product. Polymerization is through a

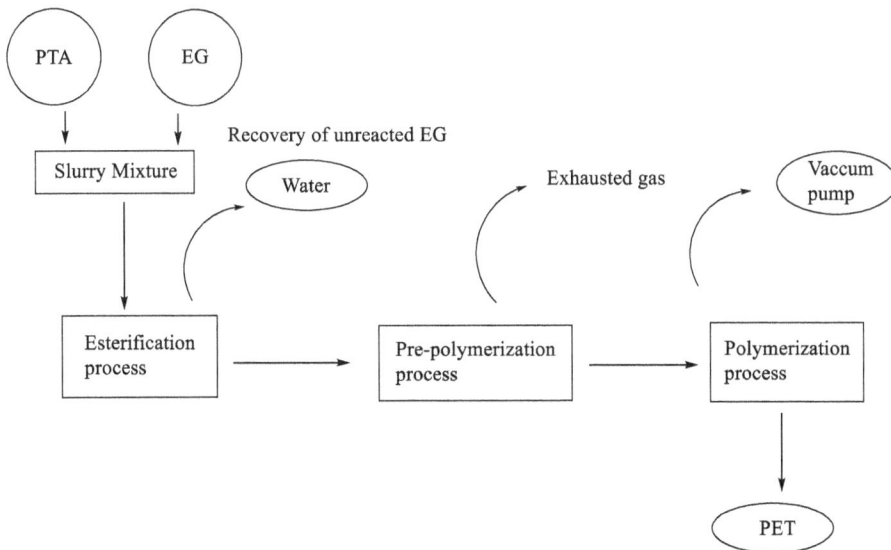

Figure 2.17: Manufacturing process of PET.

polycondensation reaction of monomers (done immediately after esterification/transesterification) with water as the by-product (Figure 2.17).

2.4.1 Types of PET

PET homopolymer and copolymer grades are available in the market. Copolymers are produced by introducing monomers that interfere the crystallinity. In PETG or PET-G copolymers, cyclohexane dimethanol (CHDM) is added to the polymer chain instead of ethylene glycol. Larger CHDM building block has six additional carbon atoms than ethylene glycol which does not fit properly with the neighbouring chains. This results in changes in crystallization and lowers the melting temperature. PETG is more amorphous and can be moulded or extruded. Another way in copolymerization is to replace TPA with isophthalic acid which also interferes crystallization resulting in lower melting point.

PET is a colourless, semi-crystalline resin. Based on how it is processed, PET is very lightweight (density: 1.38 g/cm^3) and can be semi-rigid to rigid. It can be processed by rapidly cooling the molten polymer below T_g (glass transition temperature) to form an amorphous solid to form self-reinforced composite. They are combustible at high temperatures and tend to shrink away from flames and self-extinguish upon ignition. Polyester fibres have high tenacity and E-modulus as well as low water absorption and minimal shrinkage in comparison with other industrial polymers.

2.4.2 Synthesis of polyester

Polyesters are formed from a dicarboxylic acid and a diol. Benzene-1,4-dicarboxylic acid (TPA) and ethane-1,2-diol are reacted to form the monomer for polyester or PET[13–18].

Another PE is made from benzene-1,4-dicarboxylic acid and propane-1,3-diol (which replaces ethane-1,2-diol), also known as polytrimethylene terephthalate. The benzene rings in the molecular chain give them a rigid, strong structure and high melting points (over 500 K). They are manufactured as fibres, films, or other forms for various purposes, mainly in packaging.

There are various manufacturing methods for polyester. Monomer can be prepared either by direct esterification reaction or by ester interchange reaction. Benzene-1,4-dicarboxylic acid (TPA) reacts directly with ethane-1,2-diol in the first method (Figure 2.18). Acid reacts with methanol to form dimethyl ester (DMT) and then the ester reacts with ethane-1,2-diol (transesterification).

In the presence of antimony(III) oxide catalyst, polycondensation produces the polymer at a temperature of 535–575 K and low pressure (Figure 2.19).

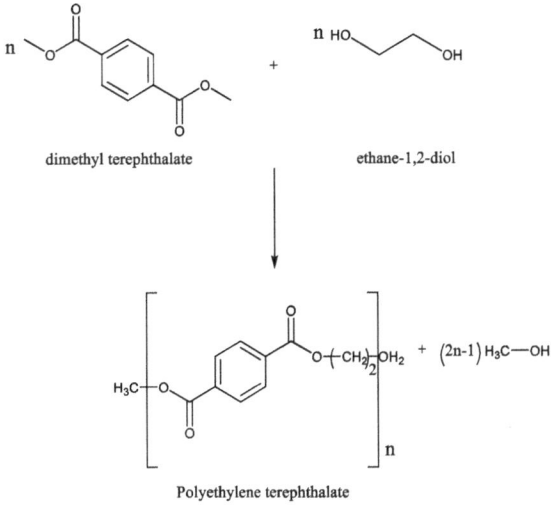

Figure 2.18: Transesterification reaction.

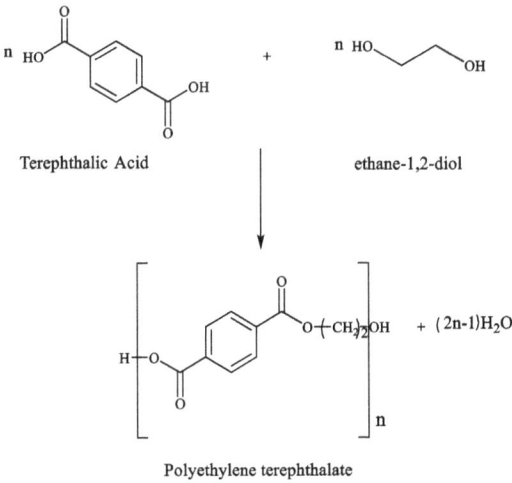

Figure 2.19: Polycondensation reaction.

Benzene-1,4-dicarboxylic acid (TPA) is manufactured by oxidation of 1,4-dimethylbenzene (commonly known as *para*-xylene) passing air into the liquid hydrocarbon dissolved in ethanoic acid under pressure, in the presence of cobalt (ll) and manganese(ll) salts as catalysts, at about 500 K.

2.5 Polymethylmethacrylate (PMMA)

Polymethylmethacrylate (PMMA) is one of the commonly used 100% recyclable thermoplastic from acrylic family of polymers. It possesses better tensile and flexural strength, transparency, and UV tolerance compared to polycarbonates but less impact strength. PMMA is generally amorphous due to the bulky side groups in the molecule. Though they are resistant to many chemicals, PMMA is soluble in esters, chlorinated hydrocarbons, and ketones. PMMA is widely used as a substitute to glass in many decorative applications, automobile parts, signboards, and so on (Figure 2.20).

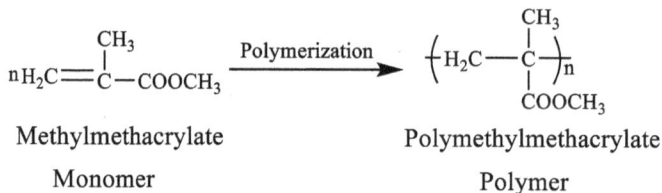

Methylmethacrylate Monomer → Polymethylmethacrylate Polymer

Figure 2.20: Schematic diagram of PMMA.

They are available in the form of pellets, sheets, or granules. They can be processed by injection moulding, compression moulding, or extrusion methods.

2.5.1 Synthesis

PMMA is produced by radical polymerization from methyl methacrylate monomer. Monomer is produced from acetone heating the cyanohydrin with 98% of sulfuric acid to form methacrylamide sulfate which is subjected to esterification to form methyl methacrylate. Detailed process of polymerization is shown in Figure 2.21.

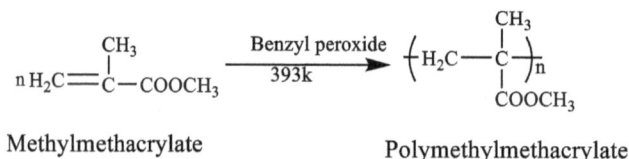

Methylmethacrylate → Polymethylmethacrylate

Figure 2.21: Synthesis of PMMA.

2.6 Polylactic acid or polylactide (PLA)

Polylactic acid (PLA) is one of the most used biopolymers derived from natural re-sources like corn, sugar cane, and cassava. PLA is a thermoplastic widely used as a 3D printing material. Though self-reinforced PLAs are developed recently, 3D-printed self-reinforced PLA composites are yet to be explored.

Basic mechanical properties of PLA are between that of polystyrene and PET. Poly-D-lactide (PDLA), poly-DL-lactide (PDLLA), and poly-L-lactide (PLLA) (regular and racemic) are the types of PLA which can be considered for self-reinforcing. Melting temperature of PLLA is 50–60 °C higher than that of PDLA. PLA is a widely used biomedical material. PLA is UV resistant and chemical resistant and possesses high overall mechanical strength (Figure 2.22).

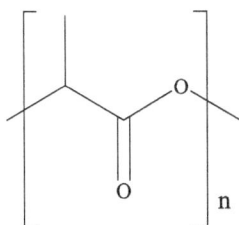

Figure 2.22: Structure of PLA.

PLA is produced by the fermentation process. Extracted starch from the source ma-terial is mixed with acid or enzymes and then heated to form dextrose. Fermenta-tion of glucose produces L-lactic acid. PLA is produced by direct polymerization or indirect condensation routes (Figure 2.23).

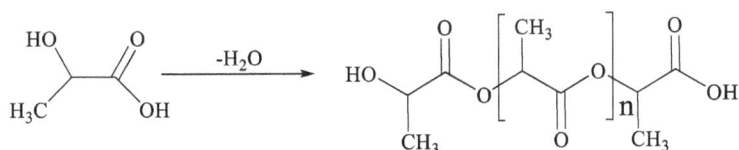

Figure 2.23: Synthesis of PLA.

Other polymers like polycarbonate, polystyrene, and acrylonitrile butadiene styrene are least explored in the development of self-reinforced composites so far due to difficulties in processability.

Table 2.1 summarizes various mechanical properties of raw materials which will be helpful in material selection.

Table 2.1: Mechanical properties of raw materials.

Material	Formula	Abrasive resistance – ASTM D1044 (mg/1,000 cycles)	Coefficient of friction	Compressive modulus (GPa)	Compressive strength (MPa)	Elongation at break (%)	Hardness – Rockwell	Izod impact strength (J/m)	Poisson's ratio	Tear strength (N/mm)	Tensile modulus (GPa)	Tensile strength (MPa)
Cellulose acetate Butyrate	CAB	–	–	–	–	60	99	260	–	–	0.3–2.0	20–60
Ethylene-chlorotrifluoroethylene copolymer	E-CTFE	5	0.07–0.08	–	–	200	R95	<1,000	–	–	1.4–1.6	48
Ethylene–tetrafluoroethylene copolymer	ETFE	–	–	–	–	250–350	R50	>1,000	–	–	0.8	28–48
Fluorinated ethylene Propylene copolymer	FEP	–	0.27–0.67	–	–	150–300	R25-45, 60 Shore D	No Break	0.48	–	0.5–0.6	14–30
Polyacrylonitrile butadiene styrene	ABS	–	0.5	–	–	45	R100–110	200-400	0.35	–	2.1–2.4	41–45
Polyamide – nylon 6	PA 6	5	0.2–0.3	–	–	–	M82	30–250	0.39	–	2.6–3.0	78
Polyamide – nylon 6, 6	PA 6,6	3–5	0.2–0.3	–	–	–	M89	40–110	0.41	–	3.3	82
Polyamide – nylon 6, 6 - 30% carbon fibre Reinforced	PA 6, 6–30% CFR	–	–	–	–	2	–	–	–	–	24	260
Polyamide – nylon 6, 6–30% glass fibre reinforced	PA 6,6 30% GFR	–	–	–	–	5	M100	120	–	–	10–11	160–210
Polyamide – nylon 12	PA 12	–	–	–	–	290–300	R84-107	–	–	–	–	50–55
Polyamide/imide	PAI	–	–	–	170–220	7–15	E72-86	60–140-notched	0.38	–	4.5–6.8	110–190

Material	Abbr.											
Polybenzimidazole	PBI	-	0.19–0.27	6.2	400	3	K115	590 unnotched	0.34	-	5.9	160
Polybutylene terephthalate	PBT	-	-	-	-	250	M70	60	-	-	2	50
Polycarbonate	PC	10–15	0.31	-	>80	100–150	M70	600–850	0.37	-	2.3–2.4	55–75
Polycarbonate–30% carbon fibre reinforced	PC – 30% CFR	-	-	-	-	2	-	-	-	-	18	170
Polycarbonate–30% glass fibre filled	PC – 30% GFR	-	-	-	-	3	-	-	-	-	5.8	70
Polycarbonate-conductive	PC	-	-	-	-	5–9	-	-	-	-	~3.8	60–70
Polychlorotrifluoroethylene	PCTFE	-	-	-	9–12	80–250	R75-112	D75-90-Shore	267	-	-	1.3–1.8
Polyetheretherketone	PEEK	-	0.18	-	-	50	M99	85	0.4	-	3.7–4.0	70–100
Polyetherimide	PEI	10	-	2.9	140	60	R125	50	0.44	-	2.9	85
Polyethersulfone	PES	6	-	-	-	40–80	M88	85	0.4	-	2.4–2.6	70–95
Polyethylene–high density	HDPE	-	0.29	-	-	-	D60-73 - Shore	20–210	0.46	-	0.5–1.2	15–40
Polyethylene – low density	LDPE	-	-	-	-	400	D41-46 - Shore	>1000	-	-	0.1–0.3	5–25
Polyethylene – UHMW	UHMW PE	-	0.1–0.2	-	-	500	R50-70	>1000	0.46	•	0.2–1.2	20–40
Polyethylene terephthalate	Polyester, PET, PETP	-	0.2–0.4	-	-	-	M94-101	13–35	0.37–0.44 (oriented)	•	2–4	80, for biaxial film 190–260
Polyimide	PI	-	0.42	-	-	8–70	E52-99	80	-	-	2.0–3.0	70–150
Polymethylmethacrylate	PMMA, acrylic	-	-	-	-	2.5–4	M92-100	16–32	0.35–0.4	-	2.4–3.3	80
Polymethylpentene	TPX®	-	-	-	-	15	R85	49	-	-	1.5	25.5
Polyoxymethylene – copolymer	Acetal – copolymer POMC	-	-	-	-	15–40	M80	70–80	0.35	-	2.3–2.8	60–70
Polyoxymethylene – homopolymer	Acetal – homopolymer POMH	-	0.2–0.35	-	-	40–75	M94	75–130	0.35	-	2.9–3.1	70

(continued)

Table 2.1 (continued)

Material	Formula	Abrasive resistance – ASTM D1044 (mg/1,000 cycles)	Coefficient of friction	Compressive modulus (GPa)	Compressive strength (MPa)	Elongation at break (%)	Hardness – Rockwell	Izod impact strength (J/m)	Poisson's ratio	Tear strength (N/mm)	Tensile modulus (GPa)	Tensile strength (MPa)
Polyphenyleneoxide	PPO (modified), PPE (modified)	20	0.35	–	–	50	M78/R115	200	0.38	–	2.5	55–65
Polyphenyleneoxide (modified), 30% glass fibre reinforced	PPO 30% GFR	35	–	–	–	2–3	L108	80	0.27	.	8–9	100–120
Polyphenylenesulfide – 40% glass fibre reinforced	PPS – 40% GFR	–	–	–	–	1.2	R123	75–80	–	–	7.6–12.0	124–160
Polyphenylsulfone	PPSu	–	–	–	–	30	M80	–	–	–	2.5	76
Polypropylene	PP	13–16	0.1–0.3	–	–	150–300, for biaxial film >50	R80-100	20–100	–	–	0.9–1.5, for biaxial film 2.2–4.2	25–40, for biaxial film 130–300
Polystyrene	PS	–	–	–	–	1.6	M60-90	19–24	0.35	–	2.3–4.1	30–100
Polystyrene – conductive	High impact conductive polystyrene	–	–	–	–	36	–	no break	–	–	1.6	27
Polystyrene – cross-linked	PS–X-linked	60–100	–	–	–	3–5	R110-120	–	–	–	1.65	55–70
Polysulfone	PSu	–	–	–	–	50–100	M91	69	–	–	2.48	70
Polytetrafluoroethylene	PTFE	–	0.05–0.2	–	–	400	D50-55-Shore	160	0.46	–	0.3–0.8	10–40
Polytetrafluoroethylene filled with glass	PTFE 25% GF	–	0.08–0.10	–	–	100–300	D60-70-Shore	144	–	–	1.7	7–20

Polyvinylchloride – unplasticized	UPVC	–	–	–	–	60	R106-120	20–1000	–	–	2.5–4.0	25–70
Polyvinylfluoride	PVF	–	–	–	90–250	D80-Shore	180	0.4	130–200 (initial)	2.1–2.6	55–110	
Polyvinylidenefluoride	PVDF	24	0.2–0.4	–	50	R77-83	120–320	0.34	–	1.0–3.0	25–60	
Silicone elastomer	MQ/VNQ/ PMQ/P VMQ	–	–	–	–	60 degrees Shore A	–	–	–	–	6.5	
Tetrafluoroethylene-perfluoro(alkoxy vinyl ether) – copolymer	PFA, Teflon PFA	–	–	–	300	–	–	–	–	–	25	

References

[1]   H. L. Stein, "Engineered Materials Handbook", Vol. 2, 167, 1998.

[2]   S. M. Kurtz, "UHMWPE Biomaterials Handbook", 3rd Edition, 3.

[3]   "Market Study: Polyethylene – HDPE", *Ceresana Research, May 2012.*

[4]   "A Guide to IUPAC Nomenclature of Organic Compounds (Recommendations 1993) IUPAC, Commission on Nomenclature of Organic Chemistry". *Blackwell Scientific Publications, 1993.*

[5]   D. Tripathi, "Practical Guide to Polypropylene", Shrewsbury: RAPRA Technology, 2001.

[6]   "Polypropylene Plastic Materials & Fibers by Porex", *www.porex.com.*

[7]   C. Maier and T. Calafut, "Polypropylene: The Definitive User's Guide and Databook", In: W. Andrew, Editor, 14, 1998.

[8]   E. P. Moore, "Polypropylene Handbook. Polymerization, Characterization, Properties, Processing, Applications", New York: Hanser Publishers, 1996, ISBN 1569902089.

[9]   "Biaxially Oriented Polypropylene Films", Granwell, Retrieved: 2012-05-31.

[10] "Materials/Polyamide". PAFA. Packaging and Film Association. *Retrieved* 19 April *2015.*

[11] A. J. Wolfe, "Nylon: A revolution in textiles", *Chem. Heritage Mag.*, 26, 3, 2008. *Retrieved 30 November 2016.*

[12] J. Clark. *"Polyamides"*, Chemguide, *Retrieved 27 January 2015.*

[13] History of Nylon US Patent 2,130,523. "Linear polyamides suitable for spinning into strong pliable fibers", U.S. Patent 2,130,947 'Diamine dicarboxylic acid salt' and U.S. Patent 2,130,948 'Synthetic fibers', all issued September 20, 1938.

[14] "Nylon 6/6 – Commercial Grades and Properties", *Emco Industrial Plastics, Inc*, Retrieved November 17, 2014.

[15] L. N. Ji, "Study on preparation process and properties of polyethylene terephthalate (PET)", *Appl. Mech. Mater.*, 312, 2013, 406–410. doi:10.4028/www.scientific.net/AMM.312.406.

[16] S. Fakirov ed., "Handbook of Thermoplastic Polyesters", Weinheim: Wiley-VCH, 1223 ff, 2002, ISBN 3-527-30113-5.

[17] A. J. Pennings, R. J. van der Hooft, A. R. Postema, W. Hoogsteen and G. ten Brinke, *Polym. Bull*, 16, 2–3, 1986, 167.

[18] R. Hoff and R. T. Mathers, "Handbook of Transition Metal Polymerization Catalyst", John Wiley & Sons, 2010.

Chapter 3
Polymer self-reinforced composites – a review

A brief review of various researches conducted on self-reinforced polymer composites (SRPCs) are presented in this chapter.

3.1 PE-based SRC

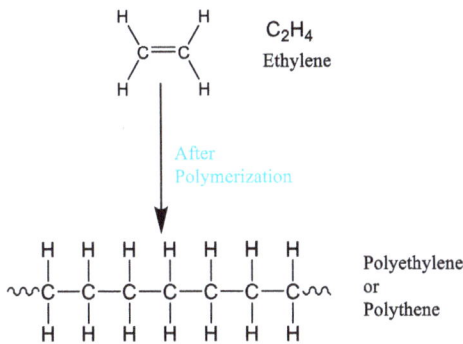

Figure 3.1: Schematic structure of polythene.

Figure 3.1 represents a schematic diagram of polyethylene (PE). The SRPC concept itself was first described through PE self-reinforced composites (SRC) [1], in which single polymer composites fabricated from PE powder and PE filaments of different melting points were analysed. Major observation made from that study is the superiority of PESRCs in interfacial shear strength measured by a pull-out test. PESRCs in this study have been prepared utilizing a difference in melting points between the components. Aligned and extended chains provide thermodynamically more stable crystals. Such crystals will have higher melting point than conventionally crystallized materials. While processing, trans-crystalline regions were grown in the matrix and a fractional melting occurs among the fibre and the matrix since the melting points are close. This results in the development of a strong interface bond with a gradient in morphologies. However, they had also observed that the interfacial strength in the PESRC was due mainly to the unique epitaxial bonding rather than the radial forces from compressive shrinkage.

Since then, it took around 17 years to report another notable work published in PESRCs [2]. Dyneema SK60 PE fibres were reinforced in a high-density PE (HDPE) matrix after rigorous selection process based on differential scanning calorimetry (DSC) thermogram data of various polyolefins. They concluded that the chemical compatibility and processing temperature are the key factors in the manufacturing

https://doi.org/10.1515/9783110647334-003

of PESRCs. Specific mechanical properties of PE/PESRCs were found promising in the fibre direction at various temperatures. They also observed a growth of transcrystalline region in the fibre–matrix interface and proposed PESRCs for closely observed applications. Research carried out to develop an alternative for the antiballistic market resulted in the development of self-reinforced HDPE composites (KAYPLA™) which are also used in panels for caravans and vans.

Three years later, the influence of oscillating pressure on the structure and properties of HDPE SRCs prepared by oscillating pressure injection moulding under low pressure were reported [3]. A highly improved tensile strength of 93 MPa and an elastic modulus of 5 GPa were reported to be due to a shish-kebab crystalline structure orientation of molecular chains along the flow direction and a more perfect crystal formation. They also studied the effect of mould temperature on mechanical performance and microstructure of HDPE SRCs prepared by melt deformation in oscillating stress field [4] and observed improved tensile strength and modulus due to the formation of 20% of the crystalline phase with shish-kebab crystalline structure under oscillating stress field. Young's modulus increased from 1 to 3.5 GPa, while yield strength of HDPE increased from 23 to 87 MPa. Double peak in DSC thermograms were an indication of the presence of spherulites (low-temperature peak) and shish-kebab crystal structures (high-temperature peak). Additionally, wide-angle X-ray diffraction (XRD) measurements showed that the molecular chains are oriented in a preferred direction combined with the above-mentioned reasons which resulted in improved mechanical properties.

Ultra-high-molecular-weight PE (UHMWPE) was developed in the early 1970s, and a complete self-reinforced UHMWPE composite was first developed and characterized in the late 1990s [5] with a focus on biomedical applications. They observed that the tensile strength, tensile modulus, and creep resistance were significantly improved when UHMWPE fibres were reinforced in an UHMWPE matrix. The longitudinal tensile strength was found to increase with fibre content though the transverse strength remained unchanged for fibre content less than 7%. The fibre content and the longitudinal tensile strength were observed to possess a linear relationship. The double notch impact strength was also found to be improved compared to plain UHMWPEs. For the first time in the historical development of SRCs, fractographic features of samples failed under a tensile load were also documented by using a scanning electron microscope (SEM). Based on the observations, they proposed that these materials are good candidates for load-bearing biomedical applications.

UHMWPE/LDPE (low-density PE) composites were first manufactured by a solution impregnation method followed by preparation of prepregs and hot compaction, in 1998 [6]. Unidirectional composites were prepared by a novel manufacturing method of using LDPE/xylene solution for crystallization of matrix solution on the surface of fibres for better impregnation at three different temperatures, and the influence of processing temperature on their mechanical properties was compared. Tensile properties were found to be improved but compressive strength was found to be very low in this study. High toughness and biocompatibility of these materials indicate their suitability of applications in ballistic devices and artificial implants.

Another interesting outcome was reported from the above-mentioned studies of Capiati, Porter, and Lacroix. Composite laminates were processed under pressurized conditions. Processing temperature for matrix was slightly higher than the melting point observed from DSC. They found that the melting point of polyolefins increases under these constrained conditions which helped in obtaining a larger processing window. This could be due to the higher enthalpy and lower entropy due to fewer molecular conformations in the amorphous regions. Thus the final properties of SRPCs will depend on how the structure changes during the compaction process.

The influence of cooling rate, processing temperature, pressure, and duration of processing on elastic modulus of HDPE SRCs was reported in 2001 [7]. An optimum processing temperature, pressure, and duration were found to be matrix dependent in this study. For the Spectra® HDPE fibre, processing above 130 °C reduces the fibre modulus and the optimum processing temperature increases as the polymer melt viscosity increases. Fibre modulus was found to be decreasing with an increase in processing pressure, and an optimum processing duration of 2 h was proposed for this grade of HDPE. It was also observed that a slower cooling rate improved the composite modulus.

Investigation of damage mechanisms, which is one of the key features of this book, was found to be reported for the first time with PESRCs in 2006 [8]. They analysed the damage mechanisms in UHMWPE/HDPE composites under monotonic tensile loading by acoustic emission (AE) technique, and the fracture surface analysis was performed with SEM. A correlation with growth of damage mechanisms and AE results proposed in this study can be further used for monitoring failure growth in UHMWPE/HDPE composites. This study also established the relevance of AE technique for failure analysis.

In a study on manufacturing methods of PE and PP (polypropylene) SRCs, a combination method of film stacking and hot compaction was compared with traditional film stacking and hot compaction methods [9]. In this method, unlike the traditional hot compaction method, the case in which fibre surface also melts along with the matrix was also studied. It was observed that this causes better wetting compared to a more traditional method. It was observed that the use of interleaved films improved the mechanical properties and provided a wider processing window compared to traditional methods of film stacking and hot compaction.

Polymer composites made from UHMWPE are promising candidates for artificial joint replacement these days. For such an application, Huang et al. presented an approach to simultaneously improve the mechanical properties, fatigue, and wear resistance of radiation cross-linked UHMWPE-based SRC of low-molecular-weight PE blend prepared by injection moulding [10]. Ultimate strength, impact strength, fatigue strength, and wear resistance were found to be improved compared to compression-moulded UHMWPE. This was attributed to the improvement in wt% of

UHMWPE by the modified processing technique. They concluded that the self-reinforced PE blend reduces the risk of failure and improves the reliability in such applications.

3.2 PP-based SRC

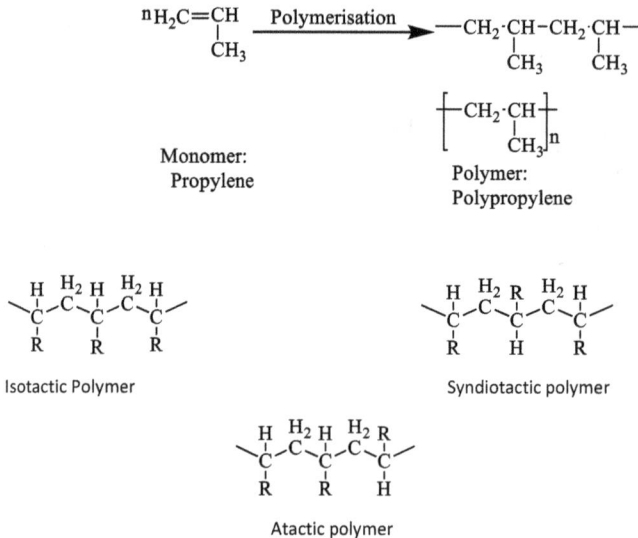

Figure 3.2: A schematic structure of polypropylene.

Figure 3.2 represents a schematic diagram of PP. Most of the polymer SRC works were reported in PPSRCs. Isotactic PP (iPP) does not possess enough strength to be used in structural applications. This drawback was addressed by self-reinforcement and that was the first reported work in PPSRCs. PPSRC was first reported in a published literature in 1995 [3]. PPSRC was prepared by oscillating pressure injection moulding under low pressure. Studies were conducted to examine the effects of processing parameters on the mechanical properties. They reported that the mechanical properties were greatly enhanced due to the presence of spherulites and shish-kebab structure and the molecular chain orientation. A second reported attempt was to prepare PPSRC by melt-flow-induced crystallization. They reported that the melt temperature is an important deciding factor in the mechanical properties of PPSRCs [11].

Hine et al. studied various parameters that influenced the hot compaction of PPSRC with five different commercially available woven fabric reinforcements [12]. Influence of various parameters such as nature of weave (fine or coarse), molecular weight and morphology, and shape of reinforcement (rectangular tape or circular fibre) on the mechanical properties was studied in this work. Major aspect of this study was the recognition of hot compacted PP as a composite material with an oriented reinforcement phase and a melted recrystallized matrix phase, and the influence of ductility in controlling the properties of the hot compacted composite. Higher molecular weight and lower crystallinity were found to improve ductility. Best properties were observed for flat tapes compared to woven fabric reinforcements.

Jordan et al. studied the pre- and post-hot compaction morphology of woven PP tapes and fibres [13]. Bundles of oriented tapes or fibres were selectively melted such that the outer surface of the tape and fabric melt, and while cooling, this outer layer recrystallizes to bind the entire structure together. Different weave styles processed under optimum processing conditions were studied for adhesion between the interfaces and direction of crack propagation along the interface in peeling test. It was concluded that a lower molecular weight PP exudes more material compared to high-molecular-weight PP and recrystallizes to contribute a low mechanical strength region. This technique further leads to the development of commercial PPSRC called curv®. Barkoula et al. validated the applicability of concept of overheating in drawable apolar iPP, UHMWPE SRCs, and less drawable polar PET and PA SRCs [14].

In a PhD work, Alcock presented various processing routes for PPSRCs and their mechanical properties. By using a combination method of constraining and co-extrusion, unidirectional Self reinforced polypropylene (SRPP) composites were manufactured with very high fibre volume fraction of greater than 0.9 with high tensile strength and modulus than their pure forms [16]. Temperature processing window could also be relaxed considerably, and excellent properties of tapes were maintained. Specific mechanical properties comparable with that of glass fibre-reinforced composites could be achieved. Another study by the same researchers reported direct forming routes for fabricating simple composite structures of PPSRCs directly from fabric without the use of pre-consolidated sheets [17]. This direct stamping method also provided better temperature processing window. Izer [18] submitted a PhD thesis which included the development and investigation of SRPP composites by utilizing the allotropes of PP. He studied the static and dynamic behaviour of PPSRCs, fracture and failure aspects, and creep behaviour of PPSRCs [18].

In another study, by exploiting the allotropy of PP, PPSRCs were prepared by widening the process window [19]. Three types of iPP matrices (α, copolymer, and β) reinforced with plain weave fabric with highly stretched tapes were used for this purpose. Consolidation and transcrystallization were observed to be improved with an increase in the processing temperature along with tensile and flexural strength and modulus. Energy absorption ability was found to be decreasing with increasing process temperature. β-Component-based PPSRCs were performing well in impact properties, and an optimum process temperature of 20–25 °C above the DSC melting temperature data provided overall and better properties. In a static and dynamic mechanical property analysis conducted on the same system of materials, film-stacked

and hot-pressed laminates at temperatures 5 and 15 °C above the melting temperature of matrix were studied. Stiffness and strength were found to improve with enhanced consolidation under increased processing temperature. Both the static and
dynamic toughness were found to be reduced with an increase in the consolidation
temperature [20].

PPSRC systems were developed from plastic waste with an environment-friendly
objective to recycle the plastic waste in accordance with the European legislation. In
that aspect, environment friendliness and recyclability were considered to be more
important than economic aspects. Thus, PPSRC was proposed to be the best alternative to glass fibre-reinforced polymer composites though the glass reinforcements
provided better strength and rigidity [21]. Flame extinguishing behaviour was also
found to be promising in PPSRCs when they were added with phosphorus-based intu-
mescent flame-retardant additives.

Another important study conducted on SRCs was about the influence of process
parameters on the properties of PPSRCs [22]. Influence of processing temperature,
dwell time, and effect of introduction of interleaved films on tensile and impact
properties were analysed in this study. Higher processing temperature was found to
be inducing more molecular relaxation and better bonding and interface. This is
found to reduce 0° tensile strength and penetration impact strength but improve
the 45° tensile strength and non-penetration impact resistance. Interleaved films
had similar influence while the dwell time had least influence on these properties.

Swolfs and his team have done some work on PPSRCs especially in impact properties. Swolfs with another group also compared the properties of PPSRCs at and
below the room temperature [23]. Stiffness was found to be improved at low temperature, and failure strain was reduced. Matrix embrittlement due to glass transition was
also reported. Below room temperature, after the non-penetration impact, the damage area was reported to be significantly reduced because of the change in damage
mechanism as the material had undergone glass transition.

Swolfs and co-workers were involved in another penetration impact study of
PPSRC [24]. In penetration impact of ductile materials, sample dimension and compaction temperature play an important role. In PPSRCs, the samples were subjected
to unwanted mechanisms of wrinkling and necking due to the influence of clamping. Energy absorbed by the material could be significantly high if this effect could
be avoided by increasing the sample size. Lower compaction temperature causes
low interlayer bonding and seriously increases this effect. They suggested increasing
the sample dimensions if wrinkles were observed.

PPSRC fabricated by hot pressing of commercially available PURE tapes were
tested and studied for their drop weight impact strength at three different energy levels
[25]. They reported a ductile behaviour and extended plasticity without a crack tip by
examining the load–deflection plots and damage mechanisms. The thickness of the
samples was high and they observed only rebound and penetration. There was no perforation noticed. Damage mechanisms observed were yarn yielding, fibre fracture, and

delamination. Tests were often disturbed by slip of the samples from the clamping de-vise due to shrinkage phenomenon. They also analysed the velocity, maximum load, and energy absorption characteristics in correlation with the structural properties.

Conventional methods of fabricating PPSRCs were not much suitable for mass production. Maintaining uniform thickness in stamping process is also a challenge. These issues were addressed with a mould-based fabrication procedure, and vari-ous mechanical properties were found to be maintained [26].

Strain rate sensitivity and stress relaxation on three-point bending properties of commercially available PPSRC curv® were also studied and reported recently [27]. Flexural strength and modulus were found to improve with increasing strain rate sensitivity. A decrease of stress with time was observed in the stress relaxation study, and Kohlrausch–Williams–Watts model was found to be more suitable to fit the data obtained from stress relaxation tests compared to the Maxwell model. They also conducted a similar type of analysis on hybrid carbon fibre/SRPP composites [28]. Bending modulus was found to be insensitive to the hybrid SRPP composites, and stress was found to decrease with time.

PPSRC was fabricated with different consolidation temperatures, and an opti-mum consolidation temperature for obtaining improved mechanical properties was proposed in a study reported this year [29]. Crystal structure and crystallinity were confirmed with the help of XRD and DSC analyses. Tensile strength was highest at 160 °C and reduced when the processing temperature was increased to 170 °C as the fibres started melting at that temperature due to high fibre shrinkage ratio. Impact strength also had the same effect with change in consolidation temperature.

3.3 PA-based SRC

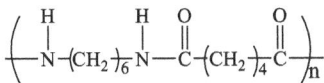

Nylon-66

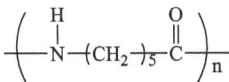

Nylon-6

Figure 3.3: Schematic structure of PA6 and PA66.

Figure 3.3 represents a schematic diagram of polyamide. Scanty literature is avail-able in polyamide-based SRCs (PASRC). First notable and reported PASRC was pre-pared from PA6 high tenacity yarn reinforcement and PA6 sheet prepared by melt quenching by exploiting the polymorphism of PA6 [30]. Film stacking followed by hot compaction was the method used for this analysis. They reported that the resulting PASRC had improved its tensile modulus by 200%, and ultimate tensile strength by 300–400% as compared with isotropic matrix film. Interfacial adhesion was improved by introducing a trans-reaction catalyst Sb_2O_3. A commercial textile yarn of PA66 was reinforced in compression moulded matrix of PA66, and mechanical properties were studied by the same team again by exploring the allotropy of PA66 [31]. They reported that this method of exploiting the allotropy provides a comfortable processing window for fabrication.

PA6 SRC prepared via in situ anionic polymerization of ε-caprolactum at a different processing temperature was studied by Gong et al. [32]. It was observed that an optimum processing temperature of 160 °C provided better mechanical properties of tension and three point bending loadings. This was observed to be due to a high reaction degree, low void fraction, and proper adhesion of interface.

A temperature of 176 °C was found to be optimum for a hybrid carbon fibre/PA12 (oriented multi-filaments) composite with a carbon fibre volume fraction of 22% [33]. For braided cloth reinforcement, a slightly higher temperature of 178 °C was found to be ideal. In short, they reported that the arrangement and size of the hybrid tows were influencing the optimum temperature requirement for processing. The brittle nature of failure was observed due to the presence of carbon fibre. They also reported in another study that the mechanical properties including the penetration impact properties were enhanced when intra-layer hot compaction method was followed with a carbon fibre volume fraction of 8% [34].

PA6 SRCs prepared from two commercial clothes were subjected to mechanical and thermal characterization by Vecchione et al. [35]. They reported an increase in flexural stiffness and a reduction in toughness with an increase in fibre content. For a thicker hybrid PA/polyurethane SRC, better mechanical property was observed. Knitted reinforced PASRC produced with a combination of powder coating and compression moulding were studied recently [36]. The influence of various parameters like reinforcement architecture, variation of fibre volume fraction, reinforcement orientation, and stacking order on tensile properties was studied in detail.

3.4 Some other popular SRCs

Self-ultra-high-strength absorbable polyglycolide-reinforced (SR-PGA) composite rods were developed for certain type of cancellous bone fracture in the 1980s. A detailed experimental study was conducted about the shear load-carrying capacity of the distal femurs of rats by implanting SR-PGA and self-reinforced poly-L-lactic acid pins by the Department of Orthopedics and Traumatology of Helsinki University Central Hospital and Tampere University of Technology, Finland, in 2001 [37].

The research in the concept of PP-based SRC was triggered by Dr. Ton Peijs in 1999 with the PURE® research team and was supported by the Dutch government by their economy, ecology, and technology programme. A patent was filed in 2001 on the concept of PPSRCs by Lankhorst Indutech with Peijs as co-inventor (WO/2003/008190; Polyolefin Film, Tape or Yarn), and the first results were published from 2003 onward. PURE® technology is licensed to the US-based Milliken and company, now marketed under the trade name Tegris® since 2006 [38].

European Union (EU) funded the HIGHBIOPOL project through the EU sixth framework programme since 2007. The objective of the HIGHBIOPOL project was to develop a range of multifunctional engineering biopolymer systems for applications in automotive, general engineering, or consumer products' markets with improved performances and properties. The main purpose of the use of biopolymers was to create biodegradable products from sustainable resources, competing with fossil-sourced polymers. HIGHBIOPOL uses a bioplastic which was made by fermenting the starches found in corn. The sugars in the corn's starch ferments into a plastic called polylactic acid (PLA) can then be used to create plastic pellets that were moulded into products. PLA-based SRCs were developed under this project [39]. Poly(methyl methacrylate) (PMMA)-based fibrre reinforced high strength and high ductility SRCs were developed in the 1990s. Later, various studies were conducted in PMMA SRPCs.

References

[1]   N. J. Capiati and R. S. Porter, "The concept of one polymer composites modelled with high density polyethylene," *J. Mater. Sci.*, 10, 10, 1671–1677, 1975.

[2]   C. Marais and P. Feillard, "Manufacturing and mechanical characterization of unidirectional composites," *October*, 45, 247–255, 1992.

[3]   Q. Guan, K. Shen, J. Ji and J. Zhu, "Structure and properties of self-reinforced polyethylene prepared by oscillating packing injection molding under low pressure," *J. Appl. Polym. Sci.*, 55, 13, 1797–1804, 1995.

[4]   Q. Guan, F. S. Lai, S. P. McCarthy, D. Chiu, X. Zhu and K. Shen, "Morphology and properties of self-reinforced high density polyethylene in oscillating stress field," *Polymer (Guildf).*, 38, 20, 5251–5253, 1997.

[5]   M. Deng and S. W. Shalaby, "Properties of self-reinforced polyethylene composites," *Biomaterials*, 18, 9, 645–655, 1997.

[6]   F. V. Lacroix, M. Werwer and K. Schulte, "Solution impregnation of polyethylene fibre/polyethylene matrix composites," *Compos. Part A Appl. Sci. Manuf.*, 29, 4, 371–376, 1998.

[7]   M. S. Amer and S. Ganapathiraju, "Effects of processing parameters on axial stiffness of self-reinforced polyethylene composites," *J. Appl. Polym. Sci.*, 81, 5, 1136–1141, 2001.

[8]   X. Zhuang and X. Yan, "Investigation of damage mechanisms in self-reinforced polyethylene composites by acoustic emission," *Compos. Sci. Technol.*, 66, 3–4, 444–449, 2006.

[9]   P. J. Hine, R. H. Olley and I. M. Ward, "The use of interleaved films for optimising the production and properties of hot compacted, self reinforced polymer composites," *Compos. Sci. Technol.*, 68, 6, 1413–1421, 2008.

[10] Y.-F. Huang *et al.*, "Self-reinforced polyethylene blend for artificial joint application," *J. Mater. Chem. B*, 2, 8, 971–980, 2014.

[11] H. X. Huang, "Self-reinforcement of polypropylene by flow-induced crystallization during continuous extrusion," *J. Appl. Polym. Sci.*, 67, 12, 2111–2118, 1998.

[12] P. J. Hine, I. M. Ward, N. D. Jordan, R. Olley and D. C. Bassett, "The hot compaction behaviour of woven oriented polypropylene fibres and tapes . I . Mechanical properties," *Polymer*, 44, 1117–1131, 2003.

[13] N. D. Jordan, D. C. Bassett, R. H. Olley, P. J. Hine and I. M. Ward, "The hot compaction behaviour of woven oriented polypropylene fibres and tapes . II . Morphology of cloths before and after compaction," *Polymer*, 44, 1133–1143, 2003.

[14] N. M. Barkoula, T. Peijs, T. Schimanski and J. Loos, "Processing of single polymer composites using the concept of constrained fibers," *Polym. Compos.*, 26, 1, 114–120, 2005.

[15] B. Alcock, *18_Single Polymer Composites Based on Polypropylene: Processing and Properties*, no. May. 2004.

[16] B. Alcock, N. O. Cabrera, N. M. Barkoula, J. Loos and T. Peijs, "The mechanical properties of unidirectional all-polypropylene composites," *Compos. Part A Appl. Sci. Manuf.*, 37, 5, 716–726, 2006.

[17] B. Alcock, N. O. Cabrera, N. M. Barkoula and T. Peijs, "Direct forming of all-polypropylene composites products from fabrics made of co-extruded tapes," *Appl. Compos. Mater.*, 16, 2, 117–134, 2009.

[18] A. Izer, "Development and investigation of self – reinforced polypropylene composites based on the polymorphism of PP," *Mech. Eng.*, 2010.

[19] A. Izer, T. Bárány and J. Varga, "Development of woven fabric reinforced all-polypropylene composites with beta nucleated homo- and copolymer matrices," *Compos. Sci. Technol.*, 69, 13, 2185–2192, 2009.

[20] T. Bárány, A. Izer and J. Karger-Kocsis, "Impact resistance of all-polypropylene composites composed of alpha and beta modifications," *Polym. Test.*, 28, 2, 176–182, 2009.

[21] K. Bocz, A. Toldy, Á. Kmetty, T. Bárány, T. Igricz and G. Marosi, "Development of flame retarded self-reinforced composites from automotive shredder plastic waste," *Polym. Degrad. Stab.*, 97, 3, 221–227, 2012.

[22] Y. Swolfs, Q. Zhang, J. Baets and I. Verpoest, "The influence of process parameters on the properties of hot compacted self-reinforced polypropylene composites," *Compos. Part A Appl. Sci. Manuf.*, 65, 38–46, 2014.

[23] Y. Swolfs, W. Van Den Fonteyne, J. Baets and I. Verpoest, "Failure behaviour of self-reinforced polypropylene at and below room temperature," *Compos. Part A Appl. Sci. Manuf.*, 65, 100–107, 2014.

[24] Y. Meerten, Y. Swolfs, J. Baets, L. Gorbatikh and I. Verpoest, "Penetration impact testing of self-reinforced composites," *Compos. Part A Appl. Sci. Manuf.*, 68, 289–295, 2015.

[25] S. Boria, A. Scattina and G. Belingardi, "Impact behavior of a fully thermoplastic composite," *Compos. Struct.*, 167, 63–75, 2017.

[26] A. Imran, S. Qi, C. Yan, D. Liu, Y. Zhu and G. Yang, "Dynamic compression response of self-reinforced polypropylene composite structures fabricated through ex-situ consolidation process," *Compos. Struct.*, 204, July, 288–300, 2018.

[27] P. N. B. Reis, L. Gorbatikh, J. Ivens and S. V. Lomov, "Viscoelastic behaviour of self-reinforced polypropylene composites under bending loads," *Procedia Struct. Integr.*, 13, 1999–2004, 2018.

[28] P. N. B. Reis, L. Gorbatikh, J. Ivens and S. V. Lomov, "Strain-rate sensitivity and stress relaxation of hybrid self-reinforced polypropylene composites under bending loads," *Compos. Struct.*, 209, August 2018, 802–810, 2019.

[29] Y. T. Hwang, S. Y. Kang, D. H. Kim and H. S. Kim, "The influence of consolidation temperature on in-plane and interlaminar mechanical properties of self-reinforced polypropylene composite," *Compos. Struct.*, 210, July 2018, 767–777, 2019.

[30] D. Bhattacharyya, P. Maitrot and S. Fakirov, "Polyamide 6 single polymer composites," *Express Polym. Lett.*, 3, 8, 525–532, 2009.

[31] M. Duhovic, P. Maitrot and S. Fakirov, "Polyamide 66 polymorphic single polymer composites," *Open Macromol. J.*, 3, 1, 37–40, 2009.

[32] Y. Gong, A. Liu and G. Yang, "Polyamide single polymer composites prepared via in situ anionic polymerization of ε-caprolactam," *Compos. Part A Appl. Sci. Manuf.*, 41, 8, 1006–1011, 2010.

[33] P. J. Hine, M. Bonner, I. M. Ward, Y. Swolfs, I. Verpoest and A. Mierzwa, "Hybrid carbon fibre/nylon 12 single polymer composites," *Compos. Part A Appl. Sci. Manuf.*, 65, 19–26, 2014.

[34] P. J. Hine, M. J. Bonner, I. M. Ward, Y. Swolfs and I. Verpoest, "The influence of the hybridisation configuration on the mechanical properties of hybrid self reinforced polyamide 12/carbon fibre composites," *Compos. Part A Appl. Sci. Manuf.*, 95, 2017.

[35] P. Vecchione, D. Acierno, M. Abbate and P. Russo, "Hot-compacted self reinforced polyamide 6 composite laminates," *Compos. Part B Eng.*, 110, 39–45, 2017.

[36] S. D. Tohidi, A. M. Rocha, N. V. Dencheva and Z. Denchev, "Single polymer laminate composites by compression molding of knitted textiles and microparticles of polyamide 6: Preparation and structure-properties relationship," *Compos. Part A Appl. Sci. Manuf.*, 109, March, 171–183, 2018.

[37] P. Nordström, "self-reinforced polyglycolide and poly-levo-lactide pins in implantation and fixation of osteotomies in cancellous bone," *An experimental study on rats*, 2001.

[38] F. Mai, E. Bilotti and T. Peijs, High performance self – reinforced polylactic acid biocomposites with degradation sensing.

[39] Á. Kmetty, T. Bárány and J. Karger-Kocsis, "Self-reinforced polymeric materials: A review," *Prog. Polym. Sci.*, 35, 10, 1288–1310, 2010.

Chapter 4
Carbon/carbon and ceramic self-reinforced composites

Polymer self-reinforced composites (SRCs) are the most developed SRCs so far. Still, carbon/carbon composites can also be treated as an SRC. Carbon/carbon composites are popular in high-temperature applications, especially in spacecrafts. Similarly very few ceramic SRCs can also be noticed in the literature. These two branches of SRCs are explained in this chapter with some reference to the existing literature.

4.1 Carbon–carbon composites

In the last few decades, carbon–carbon SRCs have advanced dramatically. Carbon is light (density 2.0 g/cm^3) and stable even at temperatures above 2,200 °C in a vacuum and inert environment. Monolithic graphite has several drawbacks, including brittleness, low strength, and difficulty fabricating complicated shapes. These limits are solved by using high-strength carbon fibres reinforced with a carbon matrix. Carbon–carbon composites are light (1.5–2.0 g/cm^3) and have excellent thermal stability, even at temperatures beyond 2,200 °C in non-oxidizing environments [1].

Organic precursor fibres such as rayon, polyacrylonitrile,, and pitch are pyrolysed in an inert atmosphere to produce carbon fibres in bulk. Carbon fibres are carbonized at 1,300 °C, whereas graphite fibres are graphitized at temperatures ranging from 1,900 to 3,000 °C. Carbon fibres have high specific strength and stiffness, high-temperature mechanical qualities, high fracture toughness, low thermal expansion, good thermal shock resistance, and machinability, among other characteristics.

According to an Indian defence update published on 6 June 2017, the Indian Army will use bullet-proof jackets made entirely of indigenous technology developed by Dr. Shantanu Bhowmik of Amrita University in Coimbatore, India, using carbon fibres, which is a collaboration between DRDO and defence ministry [2]. Carbon–carbon composites are used for the production of rocket nozzles, re-entry vehicle nose cones, leading edges, cowlings, heat shields, aircraft brakes, racing vehicle brakes, and high-temperature furnace setters and insulation.

Because of their low thermal expansion and wide range of thermal conductivities, carbon/carbon composites offer a high thermal shock resistance. At temperatures above 500 °C, however, carbon–carbon composites are highly sensitive to oxidation. The methods for protecting against oxidation include external coatings, internal oxidation inhibitors, and replacing the matrix with silicon carbide (SiC).

https://doi.org/10.1515/9783110647334-004

4.1.1 Fabrication

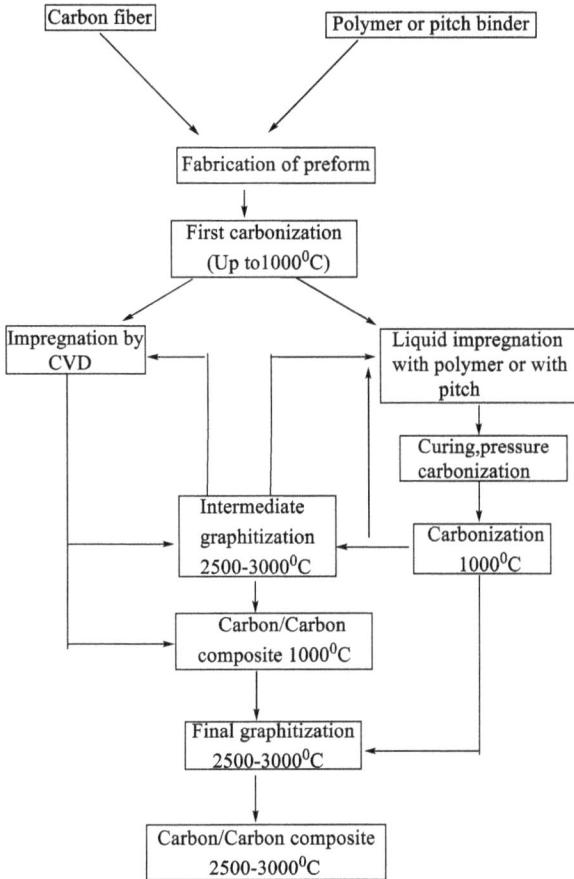

```
┌──────────────┐              ┌──────────────────────┐
│ Carbon fiber │              │ Polymer or pitch binder │
└──────────────┘              └──────────────────────┘
        ╲                    ╱
         ┌─────────────────────┐
         │ Fabrication of preform │
         └─────────────────────┘
                   │
         ┌─────────────────────┐
         │ First carbonization  │
         │   (Up to1000⁰C)      │
         └─────────────────────┘
        ╱                        ╲
┌──────────────┐          ┌──────────────────────┐
│ Impregnation by │ ◄──── │ Liquid impregnation  │
│     CVD      │          │ with polymer or with │
└──────────────┘          │        pitch         │
                          └──────────────────────┘
                                    │
                          ┌──────────────────────┐
                          │ Curing,pressure      │
                          │ carbonization        │
                          └──────────────────────┘
                                    │
         ┌─────────────────┐   ┌──────────────┐
         │ Intermediate    │   │ Carbonization│
         │ graphitization  │ ◄ │   1000⁰C     │
         │ 2500-3000⁰C     │   └──────────────┘
         └─────────────────┘
                 │
         ┌─────────────────┐
         │ Carbon/Carbon   │
         │ composite 1000⁰C│
         └─────────────────┘
                 │
         ┌─────────────────┐
         │ Final graphitization │
         │   2500-3000⁰C   │ ◄─
         └─────────────────┘
                 │
         ┌────────────────────┐
         │ Carbon/Carbon composite │
         │    2500-3000⁰C     │
         └────────────────────┘
```

Figure 4.1: Processing of carbon/ carbon composites.

Figure 4.1 depicts a schematic diagram of the carbon/carbon composite fabrication process. Carbon/carbon composites are manufactured by reinforcing braid, weave, fabric, or mat form of carbon fibre preform in the carbon matrix. The matrix could be prepared by chemical vapour infiltration, liquid infiltration, or a combination of both. Phenolic or furan resins are used for liquid infiltration. To convert the resin to carbon, it is cured and pyrolysed in an inactive environment. The process cycle is repeated to enhance the density-reducing porosity. Reinfiltration can also be done with thermoplastic carbon sources like coal tar and petroleum pitches. In one of the chemical vapour infiltration methods, carbon preform is heated to 1,000–1,200 °C and it is flooded with a hydrocarbon gas (e.g. methane, ethane, and propane). In the body of the preform, the gas pyrolyses to generate carbon. Different carbon microstructures will be formed in these two methods. The application of a single method or a mix of procedures will result in changes in matrix characteristics, which will affect the composite performance. Heat treatment at 2,000–3,000 °C yields highly active carbon–carbon composites by turning the matrix to highly crystalline graphite. This process improves various features such as the composite's modulus, load-carrying capacity, and thermal conductivity.

4.1.2 Properties

The carbon content, microstructure, and degree of graphitization has to influence over the mechanical properties of carbon fibres. The ranges of various mechanical properties are

- Ultimate tensile strengths ~ 1.38–8.5 GPa
- Tensile moduli ~ 150–830 GPa
- Thermal conductivity along the fibre's length ~ 22–1,100 W/m K
- Compressive strength ~ 100–150 MPa
- Density ~ 1.3–2.5
- Thermal shock resistance ~ 150–170 W/mm

4.1.3 Silicon carbide self-reinforced composites

The fabrication process of SiC composites is represented in Figure 4.2. Both SiC/carbon and SiC/SiC SRCs are developed so far. In SiC/C SRCs, because of their oxidation resistance, SiC matrix is employed instead of carbon matrix in the carbon fibre preform. Carbon fibres have a high-temperature capability, and ceramic matrices have a good oxidation resistance. SEP FRANCE produced C/SiC composites for liquid-propellant rocket and air-breathing engines, thrust, hot gas valves and tubes, and space aircraft thermal structures.

SiC/SiC SRCs are formed of SiC matrix and reinforcement. They can survive thermal cycling and have great oxidation resistance for lengthy periods. SiC/SiC composites are used in liquid-propellant rocket engine chambers, jet engines, gas turbine components, and space thermal structures. Still, SiC–SiC composites are observed to be thermally unstable above 1,200 °C [3].

4.1.4 Applications

Hot glass handling tools, hot press dies, and pistons are just a few of the applications for carbon/carbon composites. The majority of applications are in aerospace and defence structures. Carbon SRCs are used in a variety of defence applications, including

```
┌─────────────────────┐
│   KD-II SiC fiber   │
└─────────────────────┘
          ↓
┌─────────────────────┐
│ 4-Step 3D fiber preform│
└─────────────────────┘
          ↓
┌─────────────────────┐
│ C/Sic multi-interphase│
│      deposited      │
└─────────────────────┘
          ↓
┌─────────────────────┐           ┌──────────┐
│  PCS/xylene solution│◄──────────│          │
│ infiltration in vaccum│         │          │
└─────────────────────┘           │          │
          ↓                       │          │
┌─────────────────────┐   ┌──────────┐        │
│    Crosslinking     │   │  Repeat  │        │
└─────────────────────┘   └──────────┘        │
          ↓                    ▲              │
┌─────────────────────┐        │              │
│     Pyrolysis       │────────┴──────────────┘
└─────────────────────┘
          ↓
┌─────────────────────┐
│  SiCf/SiC composites│
└─────────────────────┘
          ↓
┌─────────────────────┐
│   SiC over-coating  │
└─────────────────────┘
```

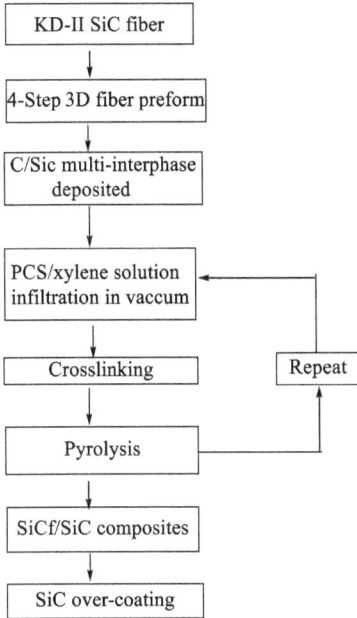

Figure 4.2: Fabrication of SiC SRCs.

limited-life missile engine components, new fighter aircraft exhaust sections, hypersonic vehicle fuselage and wing components, and space defence satellite structures.

Other applications include high-speed train and special automobile brakes and clutches, forging dies and moulding crucibles, corrosive chemical reactors and heat exchangers, fuel cells, high thermal conductivity electronic substrates, prosthetic devices, and internal combustion engine components, plasma limiters for nuclear fusion devices, and inert gas ducting and heat exchangers for gas-cooled fission reactors.

NASA created an 11.4-cm carbon-composite barrier to withstand temperatures of 1,370 °C outside the spaceship for sending robotic spacecraft to the Sun in 2018. Carbon–carbon composites are widely used for the components of military planes and missiles. Re-entry bodies, rocket nozzles, strategic missile departure cones, and military aircraft braking discs are just a few of the applications. Commercial aviation applications include brake discs for transport aircraft.

Almost 60% of all carbon–carbon composites made are utilized in aircraft brakes around the world. Because of its strong heat conductivity, and low thermal expansion coefficient, carbon–carbon composite was chosen. Carbon–carbon composite brakes have a heat capacity of around 2.5 times more than steel. When compared to other materials such as metal or organic brakes, carbon composite brakes are lighter than steel and have the potential to absorb a lot of energy in a short period. Carbon brakes have been used in commercial aircraft brake applications since the 1980s. Concorde supersonic transport was the first time user.

Carbon-based SRCs are also used in the brake systems of automobiles such as the Mercedes-Benz C215 Coupe F1 edition, as well as the Bugatti Veyron and certain current Bentleys, Ferraris, Porsches, Corvette ZR1, ZO6 and Lamborghinis, Audi cars such as the D3 S8, B7 RS4, C6 S6 and RS6, and the R8. [4].

4.2 Ceramic SRCs

Ceramics are a class of solid non-metallic, inorganic materials with metal, non-metal, or metalloid atoms in ionic and covalent bonds. Ceramics have useful features such as high strength at high temperatures, even exceeding 1,500 °C, good oxidation resistance, and great electromagnetic transparency. But their fracture toughness is very low and they fail quickly under external load. Failure due to quick crack initiation and propagation in ceramic materials can be prevented effectively by embedding reinforcements that block crack propagation. Suitable reinforcements improve their properties and make them suitable for structural applications. Particles, whiskers, fibres, and powders can all be used as reinforcements. This chapter deals mostly with aluminium and zirconium-based SRCs as the current research is mostly concentrating on these materials worldwide. Various combinations of matrices and reinforcements are represented in Table 4.1.

Table 4.1: Ceramic materials used for matrix and reinforcement (excluding carbon).

Ceramic matrix materials	Ceramic reinforcements
1. Alumina (Al_2O_3)	1. Alumina
2. Silicon carbide(SiC)	2. Silicon carbide
3. Aluminium nitride(AlN)	3. Aluminium nitride
4. Silicon nitride (Si_3N_4)	4. Silicon nitride
5. Zirconium dioxide/zirconia and partially stabilized zirconia	5. Mullite ($SiC-Al_2O_3$)

4.2.1 Ceramic self-reinforced composites

A few sorts of ceramic SRCs are
1. Silicon nitride/barium aluminosilicate (BAS) composite
2. Ca-hexaaluminate/alumina composite
3. Aluminium matrix/Al_2O_3 particles

4.2.2 Silicon nitride/barium aluminosilicate composites

Silicon nitride (Si_3N_4), a thick material with great strength and toughness for high-temperature applications, was created in the 1960s and 1970s. It is utilized in engine components, bearings, and cutting tools because of its low thermal expansion coefficient, which provides superior thermal shock resistance. Dense silicon nitride is difficult to produce. At high temperature around 1,850 °C, silicon nitride breaks down to silicon and nitrogen. Thus, silicon nitride is produced by adding some chemicals and inducing bonding indirectly. These additives are needed to make dense Si_3N_4 because they react with SiO_2 on the surface of Si_3N_4 to produce a liquid phase that aids densification. At high temperatures, the addition of oxide in Si_3N_4 leads to the formation of a grain-boundary glass phase, which softens and impairs the characteristics. BAS meets the criteria of forming a liquid phase at low temperatures that then crystallizes entirely into a compound that possesses a high melting point. These sintering aids induce a degree of liquid-phase sintering. Sintered silicon nitride, hot-pressed silicon nitride, and reaction-bonded silicon nitride are the three types of silicon nitrides. BAS ($BaAl_2Si_2O_8$) is a glass-ceramic feldspar mineral and has high chemical stability and mechanical properties. In all of the studies on BAS/Si_3N_4 composites, BAS was produced from oxide precursors such as $BaCO_3$, Al_2O_3, and SiO_2. BAS has a high melting point (1,760 °C) and a low thermal expansion coefficient (2.29×10^{-6} per degree Celsius form from 22 to 1,000 °C) [5]. Relatively small mechanical properties of BAS can be enhanced by adding whiskers, platelets, and continuous fibres. According to some research findings, rod-like Si_3N_4 can be formed in situ from Si_3N_4 in the presence of liquid BAS. The BAS/Si_3N_4 composite that developed had a high strength and fracture toughness.

In situ reinforced silicon nitride–BAS composites have mechanical properties that are comparable to standard hot-pressed dense silicon nitride ceramics, with room temperature flexural strength surpassing 900 MPa and fracture toughness approaching 7 MPa $m^{1/2}$. They are employed in a wide range of applications where high strength, toughness, low dielectric constant, low loss tangent, high thermal shock resistance, high thermal stability, low thermal expansion, and low loss cost are required. Among them is microelectronic packaging, ceramic valves, and other applications [6].

One of the methods of creating the silicon nitride/barium aluminosilicate is by pressureless sintering at 1,800 °C for 2 h. To create BAS, $BaCO_3$ was wet-milled for 24 h in anhydrous alcohol in a plastic bottle with 32 wt% SiO_2 and 27.1 wt% Al_2O_3 powder. To make BAS, the slurry was dried, and the powder mixture was sieved before being sintered at 1,300 °C for 2 h. The BAS powders were then pulverized and screened using a 150 m screen.

Rod-like Si_3N_4 seeds are made with Si_3N_4 powder (0.5 m) as starting powders, Y_2O_3 (99.9% purity) and MgO (99.9% purity) as sintering additives, and additional quantities of 5 and 2 wt%, respectively. The aforementioned initial powder mixture was wet-milled in anhydrous alcohol for 24 h in a plastic bottle using high-purity Si_3N_4 balls. The slurry was milled, dried, sieved, and heated for 1 h at 1,800 °C under 0.6 MPa nitrogen pressure in a furnace. The hot powder was crushed and then subjected to acid washing procedures to remove the remaining glassy phase.

BAS glass ceramic is a useful tool for Si_3N_4 densification and phase transition. In situ-produced rod-like Si_3N_4 whiskers can significantly improve the mechanical characteristics of BAS glass ceramic. The room temperature flexural strength and fracture toughness of 40 wt% BAS/Si_3N_4 improve by 357% and 48%, respectively, as compared to the BAS matrix alone. Extending the sintering time can improve the quantity of Si_3N_4 phase change and thus the mechanical characteristics of the material [7].

Another effective method for solidifying ceramic slurry is to use freeze gelation rather than changing the sol pH to make silicon nitride–barium aluminium silicate (Si_3N_4/BAS) composites. Slurry preparation, freeze gelation, drying, and sintering are some of the steps involved in this procedure. A consistent network structure of silica gel can be formed by freezing pure sol. After the added barium oxide and alumina absorbed silica in sol to make the $BaO-Al_2O_3-2SiO_2$ system, which was used as a sintering aid for nitride silicon, sintering was used to manufacture Si_3N_4/BAS glass ceramic matrix composites. The shrinkage rate was 0.6%, and the flexural strength, work of fracture, and density of sintered body were correspondingly 342 MPa, 181 J/m^2, and 3.0 g/cm^3, respectively [8].

4.2.3 Ca-hexaaluminate/alumina composite

Calcium hexaaluminate (CA6) is integrated into the matrix of magnesioaluminate spinel-alumina (MA-A) by infiltrating a saturated calcium acetate solution into a porous preform made of Al_2O_3 and MgO powders, resulting in a pseudo-self-reinforced aluminium compound. After sintering at high temperatures, functional CA6/(MA-A) composites with graded fracture toughness are created. One-fourth length of the preform was perpendicularly immersed in the solution. In the porous preform, the calcium acetate solution is absorbed. The MA-A region could give structural support due to its increased Vickers hardness and density. Fabrication steps of calcium aluminate are represented in Figure 4.3.

According to the microstructural analysis, MA in the burned composite is an alumina-rich non-stoichiometric spinel with a lower atomic ratio of Mg/Al and smaller lattice fringe spacing. The addition of CA6 to the MA-A matrix slowed the densification process slightly. The hardness changed very little when the distance between the infiltration end and the hardness end was small (as in CA6/(MA-A)). When the distance got vast, though, it grew with the distance and eventually levelled out. Because of its higher Vickers hardness and density, the MA-A area could help sustain the composites' structure. The presence of less hard CA6 in the CA6/(MA-A) area resulted in a lower Vickers hardness. This region had higher fracture

```
┌─────────────────┐  ┌─────────────────┐  ┌─────────────────┐  ┌─────────────────┐
│ Aqueous solution│  │ Aqueous solution│  │ Aqueous solution│  │ Solution of Eu₂O₃│
│of calcium nitrate│  │  of Aluminium   │  │  of citric acid │  │  in nitric acid │
│    Ca(NO₃)₂     │  │ nitrate Al(NO₃)₃│  │     H₄Cit       │  │                 │
└─────────────────┘  └─────────────────┘  └─────────────────┘  └─────────────────┘
```

Aqueous solution of $Ca(NO_3)_2$ + $Al(NO_3)_3$ + H_4Cit
Stirring for 1.5h

Drying at 1300C for 5h. cooling at room temperature

Anneling in air at 10000C for 3h(with heating rate of 50C/min)

Calcium Aluminate

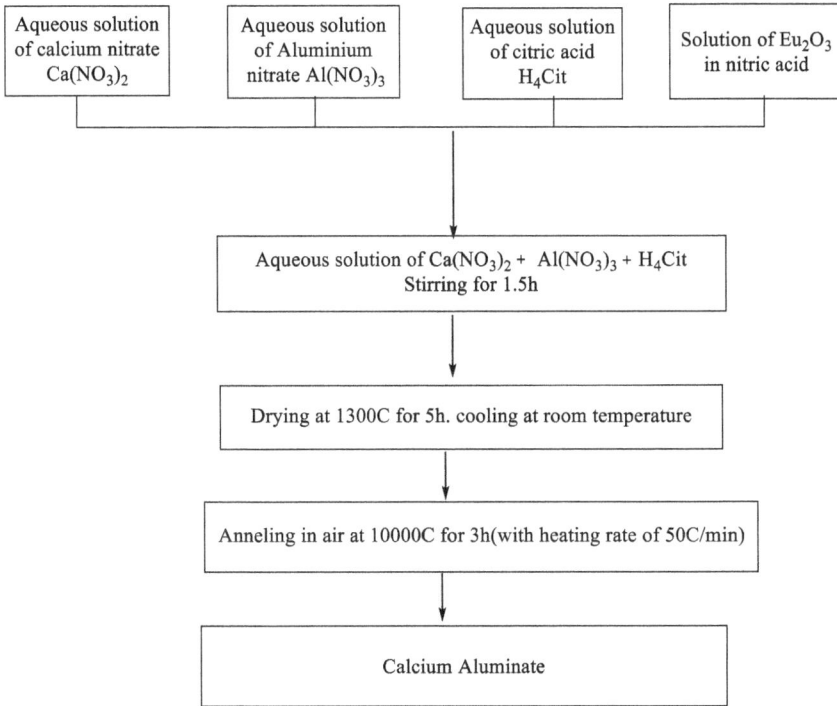

Figure 4.3: Fabrication stages of calcium aluminate.

toughness than the MA-A region, which could be due to the plate-like CA6's crack-deflection and crack-bridging effects. This area was tougher, which helped to limit crack development and increase spalling resistance. These functionally graded (CA6/(MA-A)) composites with new microstructures and characteristics could be employed in applications requiring resistance to wear, oxidation, thermal shock, fracture, chemical corrosion, and/or heat. This is a simple process that can be used to create various multifunctional functional materials [9].

4.2.4 Aluminium matrix/Al₂O₃ composite

Powder metallurgy (PM) step to fabricate calcium aluminate composites is explained in Figure 4.4. These composites are made using PM and hot extrusion of the EN AW-AlCu₄Mg1(A) aluminium alloy. To obtain a consistent dispersion of reinforcement particles in the matrix, in a laboratory vibratory ball mill, particles of the initial components were wet mixed (methanol slurry). After that, the powder combination was dried outside in the open air. In a laboratory vertical unidirectional press with a 350 kN capacity, the components were initially compacted in a cold state in a 26 mm

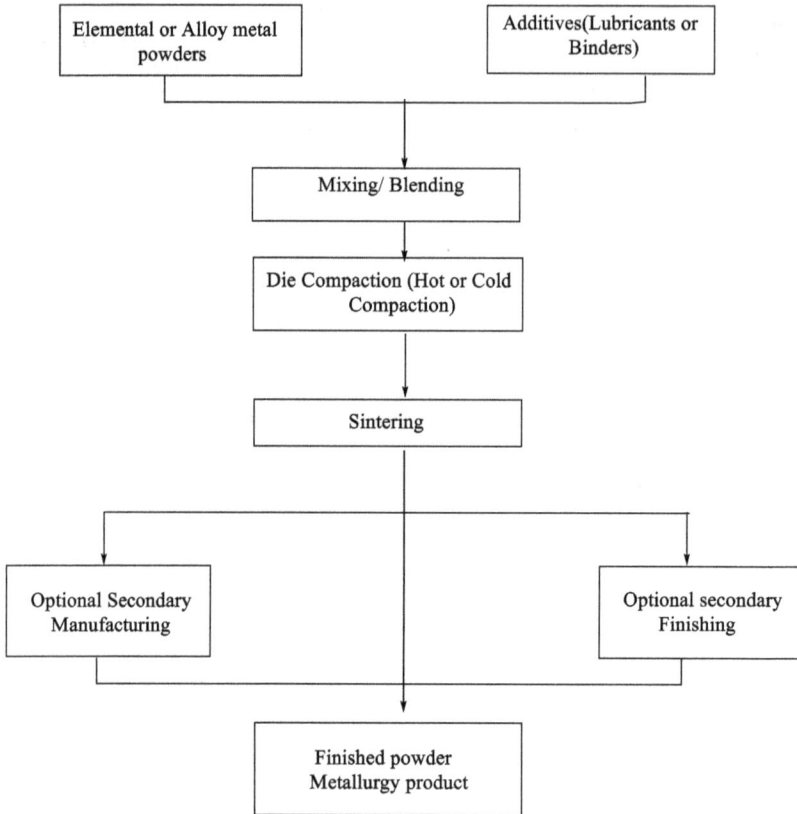

Figure 4.4: Powder metallurgy steps to produce calcium aluminate composites.

diameter die. Before being extruded at a pressure of 500 kN, the resulting PM compacts were heated to a temperature of 480–500 °C. As a result, the finished product was 8 mm diameter bars. Sintering Al_2O_3 powder with a pore creation agent in the form of carbon fibres resulted in porous ceramics. Preparation of powder and carbon fibres mixture, pressing of prepared powder combination, and compact sintering were all parts of the ceramic preform manufacturing process. Carbon fibres accounted for about 50% of the total weight. A per cent polyvinyl alcohol soluble in water was added to make pressing easier. The ceramic powder and carbon fibre combinations were uniaxially pressed in a steel mould with a 30 mm inner diameter in a hydraulic press. The highest pressure was 100 MPa, with a 15-spressing time. Compacts were sintered at a rate of 20 L/min in an air environment in a pipe furnace. The temperature for the sintering process was set to prevent carbon fibre deterioration (heating for 10 h at 800 °C) and Al_2O_3 powder sintering for 2 h at 1500 °C. The resulting ceramic performs with a porosity of 80.80% [10, 11].

In a furnace, prepared ceramic preforms were heated to a temperature of 800 °C. The graphite-covered form was heated to 450 °C (the maximum temperature of the press plates) before being filled with preform and EN AC–AlSi$_{12}$ liquid alloy at 800 °C. The stamp was used to cover the entire surface before being placed in a hydraulic plate press. The highest infiltration pressure was 100 MPa, with a load time of 120 s. After solidification, the materials were removed from the mould and allowed to cool in a pressurized air stream. Metallographic examinations of the obtained composite materials were performed on cross-sections longitudinal in the extrusion and infiltration directions of the composite materials using a light microscope to investigate the structure and determine the distribution of the reinforcing particles in the matrix [12].

Both ways of producing ceramic-enhanced aluminium matrix composite materials achieve the necessary structure and are practicable. The ability to manufacture small elements near net shape is unquestionably an advantage of PM, whereas pressure infiltration allows for the manufacture of locally reinforced elements from composite materials with good surface quality, but it needs more energy than powder metallurgy, which is the main drawback of its application.

References

[1]   T. Windhorst and G. Blount, *Mater. Des.*, 18, 1, 1997, 11.

[2]   www.Defenceupdate.com, June 8 2017.

[3]   G. R. Devi and K. R. Rao, *Def. Sci. J.*, 43, 4, 1993, 369.

[4]   E. Fitzer, *Carbon*, 25, 2, 1987, 163.

[5]   E. J. Wuchina and I. G. Talmy, In: Proceedings of the 14th conference of Metal, carbon and ceramic composites(part 1), NASA conf publ, 1990, 239.

[6]   K. W. White, F. Yu and Y. Fang, Handbook of Ceramic Composites, Springer US, 2005, 251.

[7]   F. Ye, S. Chen and M. Iwasa, *Scr. Mater.*, 48, 2003, 1433.

[8]   J. Yu, S. Li, Y. Lv, Y. Zhao and Y. Pei, *Mater. Lett.*, 147, 2015, 128

[9]   F. Ye, L. Liu, J. Zhang, M. Iwasa and C. L. Su, *Compos. Sci. Technol.*, 65, 2005, 2233.

[10] B. Wang, J. Yang, R. Guo, J. Gao and J. Yang, *J. Mater. Sci*, 44, 2009, 1351.

[11] S. Yi, Z. Huang, J. Huang, M. Fang, Y. Liu and S. Zhang, *Sci. Rep.*, 4, 2014, 4333.

[12] A. Wolderczyk-Fligier, L. A. Dobrzanski, M. Kremzer and M. Adamiak, *J. Achiev. Mater. Manuf.*, 27, 1, 2008, 99.

Chapter 5
Processing and manufacturing self-reinforced polymer composites

Selection of processing method of self-reinforced polymer composites (SRPC) depends on various factors like the material to be processed, desired properties and requirements, availability of processing methods, and cost. Processing methods like compression moulding, tape winding, film stacking, extrusion, powder impregnation, melt mixing, hot pressing are widely used for the production of SRPCs. As most of the SRPCs are formed of thermoplastic polymers, heat treatment is an important processing method for many of them. Multiple processes are used in many of the cases.

5.1 Processing conditions

As heat treatment is involved in SRPCs, the processing temperature is the major factor to be considered in processing SRPCs. Heating/cooling time, pressure to be applied, and so on are the other associated parameters. Major attention to be given to avoid the heat-associated property degradation of the reinforcement while selecting the processing temperature and related parameters. Exploring the possibilities in obtaining wide window of processing temperature benefits the processing of SRPCs. Processing will be convenient when there is a considerable difference in the softening or melting temperature of matrix and reinforcement. Still, various researchers pointed out that the narrow processing window benefits in better interface properties due to partial melting and formation of trans-crystalline layers.

Melting temperature (T_m) of polymers depends upon the crystallinity of polymers. Crystallinity is the order of alignment of molecular chain of polymers. Polymers are composed of long molecular chains of polymers from irregular, intertwined coils in the melt. Some polymers retain such a disordered structure while cooling and forms amorphous solids. Some of the polymer chains rearrange while cooling, forming partly ordered regions. Because of the intertwined coils, within the ordered regions, the polymer chains form aligned and folded structure. These regions which are neither crystalline nor amorphous are known as semi-crystalline. Isotactic polypropylene (PP), linear polyethylene (PE), and polyethylene terephthalate (PET) are some of the examples of semi-crystalline polymers. Semi-crystalline polymers are preferred over the amorphous polymers for reinforcement due to their higher stiffness and strength. Amorphous or semi-crystalline polymers may be considered for matrix.

Rate of crystallinity of a polymer depends upon various factors like tacticity, polymorphism, co-polymerization, molecular weight, and molecular architecture.

https://doi.org/10.1515/9783110647334-005

Let us examine the influence of each of these factors in crystallinity before moving to the major processing methods.

5.1.1 Tacticity

Some of the long polymer chains (e.g. PP, $(C_3H_6)_n$) contain a steriogenic carbon atom in their chains which is bonded to four different atoms. In the case of PP, alternate carbon atom in the chain carries a CH_3 group. If the CH_3 group is on the same side of the chain, PP is called isotactic, while PP (iPP, $T_m \sim 165$ °C) with CH_3 group on the alternate sides on two consecutive carbon atom in zigzag form is called syndiotactic PP (sPP, $T_m \sim 135$ °C). An atactic PP (aPP) is having the CH_3 group in irregular form (refer Chapter 2). iPP and sPP are highly crystalline due to their regular structure possessing high T_m while aPP is amorphous and melts well below the melting temperature.

5.1.2 Polymorphism

Each type of polymer may exist in more than one crystalline form even if they have same tacticity due to various packing of the chains. iPP is having four stable polymorphic forms based on their crystal unit cell parameters. Monoclinic, hexagonal, triclinic, and pseudo-hexagonal lattices in crystal units form α-, β-, γ-, and δ-forms of iPP. These crystalline modifications possess different T_m and can be well utilized for selection of matrix and reinforcement with larger range of processing temperature. Nylon 6,6 is another example which exhibits polymorphism.

5.1.3 Molecular structure

The length and density of branches also affect the crystallinity. Low-density PE (LDPE) is having less melting temperature, and as the density increases, crystallinity also improves resulting in high melting temperatures.

5.1.4 Molecular weight

Molecular weight is another deciding parameter of melting temperature. Softening or melting temperature of amorphous polymers increases with increasing molecular weight. For semi-crystalline polymers, decreasing molecular weight enhances the crystallinity. Crystallinity or melting temperature of polymers does not depend upon molecular weight alone. Various other factors have to be considered along with the molecular weight before coming to a conclusion. Ultra-high-molecular-weight PE (UHMWPE) does not contain long chains unlike the LDPE and thus possesses higher melting temperature.

5.1.5 Copolymerization

Two or more monomers polymerize to form copolymers. When different monomers polymerize, the regularity of the chain will be reduced affecting the crystallinity and thus reducing the melting temperature.

5.1.6 Heat treatment criteria

Proper control over the processing temperature is very much essential in processing SRPC. Reinforcement is susceptible to temperature-induced property degradation. Amorphous polymer matrices are processed at their softening temperature and semi-crystalline matrices are processed at their melting temperature. Reinforcement may be intact or partially molten maintaining their mechanical properties.

5.1.7 Nucleation

Heterogeneous nucleation in fibre-reinforced thermoplastics causes a crystal growth and a trans-crystalline interphase is formed. This interphase improves the mechanical properties such as strain which will be more uniformly distributed, and the resistance to compressive and radial stresses is increased. Interfacial bond strength also gets improved. Amorphous fibres do not show this kind of crystal growth. Studies on SRPCs indicate such a development of trans-crystalline layer.

5.2 Processing methods

There are various processing methods developed for SRPCs [1]. All the methods make use of temperature and pressure as major processing parameters. As they are processed by softening or melting, there must be proper releasing agents and peel ply to be used in order to avoid sticking of the final product to the surfaces. Proper control over the parameters and careful processing are very essential in manufacturing SRPCs. Hot compaction, compression moulding, injection moulding, and filament winding are some of the most widely used methods in processing SRPCs.

5.2.1 Hot compaction

This is one of the commonly used methods to manufacture SPRCs. The matrix phase of the composite is produced by softening/melting the polymer sheet, and the reinforcement is obtained by partial melting of fibres under suitable temperature and pressure resulting in excellent bonding between the matrix and fibre phases [2]. Molten outer surface of the fibres get bonded with the matrix after cooling, while remaining part of the fibres stays intact contributing the reinforcement properties. Processing temperature has to be carefully controlled in order to avoid excessive heating and insufficient heating. Excessive heating deteriorates the properties by disturbing the molecular orientation, and insufficient heating leads to a poor interfacial bonding. The properties of the polymer enhanced through controlled molecular orientation result in self-reinforcement (Figure 5.1).

PE [3–5], PET [6], PP [7], nylon 6,6 [8], and polymethyl methacrylate [9] are some of the polymers that can be self-reinforced by hot compaction. The most recent research has focused on requirement to define protocols for the hot compaction process in terms of the temperature/pressure profile required by understanding of the molecular structure which occurs as a result of the compaction process.

Figure 5.1: Hot compaction.

5.2.2 Overheating

Overheating is the processing done by heating the matrix above their melting temperatures. Overheating produces monoextrudates (identical polymers with inherently homogeneous structure) with fibres getting partially molten. The oriented fibres are embedded in the molten polymer matrix of the same grade. Polar and apolar polymers are processed by this method. Interchain interactions are weak in apolar polymers and they are highly drawable. Due to strong interchain interactions in polar polymers, they are less drawable. Draw ratio of iPP is greater than 14 and that of PET is 4. PP, PE, PET, and polyamides (nylon) can be treated by this method.

5.2.3 Film stacking

Polymer fabrics are usually reinforced with polymer sheets by using this method (Figure 5.2). Hot compaction and overheating are usually integrated with film stacking based on the requirements. Still, the method is described as a separate one as there are some other processes associated with it.

Fabrics are sandwiched between the matrix films and hot pressed to form the composite laminates. A combination of copolymer/homopolymer, polymorphic forms, or the property of tacticity can be well utilized while selecting the matrix and the reinforcement in order to provide better processing temperature window. PP, nylon, polylactic acid (PLA), PE, or any thermoplastic polymer available as films can be processed by this method. This method offers the freedom to select wide processing window and less expensive processing.

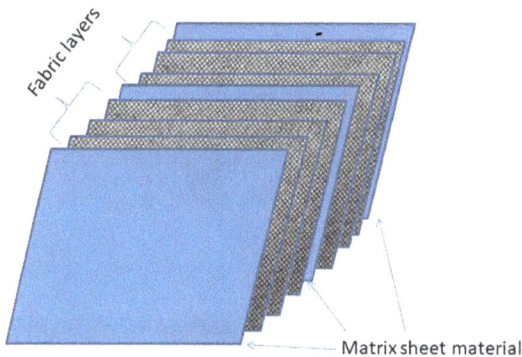

Figure 5.2: Film stacking.

While processing polymorphic forms of PP, film stacking method is followed, often accompanied, by compression moulding or tape winding. PET and PLA sheets are also self-reinforced followed by compression moulding.

5.2.4 Co-extrusion

Another main method by which SRP composites are produced is co-extrusion as shown in Figure 5.3. Co-extrusion is mainly used for the production of self-reinforced tapes with skin on outer layers with a core inner section [10]. The skin layer is composed of a copolymer with a lower melting temperature and the core is composed of a homopolymer. Polymer tapes are extruded from a high melting point grade of the polymer to be processed. A low melting point grade of the same family of polymers is extruded on the surface of the tape. These tapes are woven to form a fabric. Post-processing leads to the development of desired shapes with the outer layer of each tape melts before the inner core of oriented polymer. Under pressure, this low melt grade flows throughout the fabric. After processing by hot pressing, the skin will act as a matrix and the core acts as the reinforcement. Core layer or reinforcement is around more than 80%, and the properties of the core will be well preserved during the processing.

The tapes are constrained by the moulding pressure during consolidation to increase the melting temperature of the core material. An enlarged processing window and a high volume fraction of the reinforcement as mentioned before are obtained by this process. Some of the examples of self-reinforced PP fabrics and sheets are by Don & Low and Lankhorst.

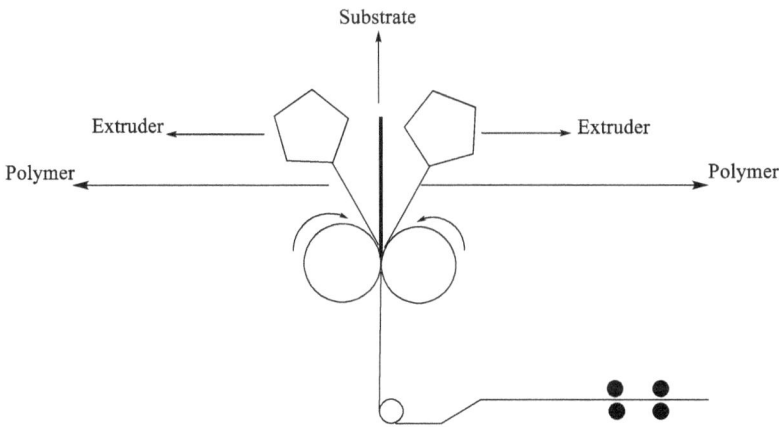

Figure 5.3: Co-extrusion.

5.2.5 Compression moulding

Compression moulding is usually associated with other methods like film stacking, injection moulding, extrusion, or winding. Most of the types of self-reinforced polymers can be processed through compression moulding.

In compression moulding, preheated moulding material is placed in an open, heated mould cavity (female die). The mould is closed with a plug member (male die), and adequate pressure is applied to maintain the material to be in proper contact with all mould areas, and heat and pressure are controlled properly. Surface hardness of the mould is improved by polishing (Figure 5.4).

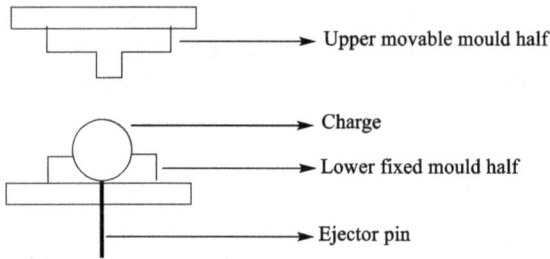

Figure 5.4: Compression moulding.

This method is suitable for moulding complex, high-strength composite thermoplastics moulded with chopped strand, unidirectional tapes, randomly oriented fibre mat, or woven fabrics. The advantage of compression moulding is its ability to mould large shapes and lowest cost compared with transfer moulding and injection moulding. It wastes relatively little material, giving it an advantage when working with expensive compounds. Drawbacks of compression moulding include poor product consistency and difficulty in controlling flashing. Slight fibre length degradation is another disadvantage.

UHMWPE powder reinforced with UHMWPE gel spun unidirectional and cross-ply fibre (10%) is processed by compression moulding at a temperature of 180 °C and at a pressure of 7 MPa. Studies revealed that the mechanical, creep, and impact properties are superior to pure UHMWPE.

PP sheets reinforced with plain woven multifilament fabric with 50% volume fraction are processed by two-step compression moulding and observed an improved tensile strength. PP sheets are melted at 200 °C with 1 MPa pressure for 10 min, and in the second compression step, fabric is sandwiched and processed at a temperature of 125–150 °C with a pressure of 9 MPa for 10 min.

5.2.6 Injection moulding

Injection moulding makes use of molten die casting method by which three-dimensional shapes can be produced while extrusion process is more suitable for two-dimensional shapes (Figure 5.5). Along with the conventional method of injection moulding, there are two another advanced methods used for the processing of SRPC systems. They are oscillating packing injection moulding (OPIM) or dynamic packing injection moulding (DPIM). Most of the thermoplastics can be processed through this method. Injection moulding is a fast, high-volume, closed moulding process used mostly for processing reinforced thermoplastics.

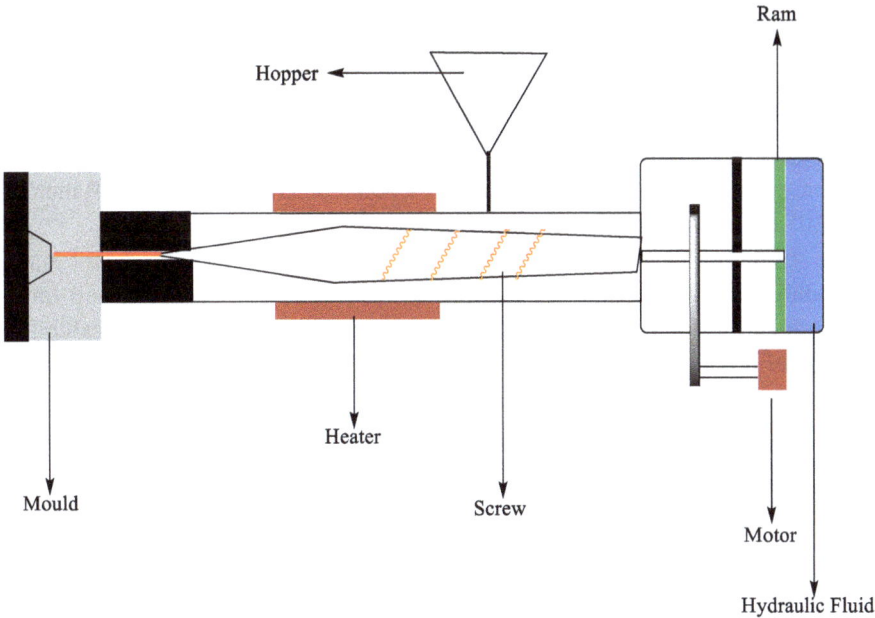

Figure 5.5: Injection moulding.

5.2.6.1 Oscillating packing injection moulding (OPIM) or dynamic packing injection moulding (DPIM)

DPIM or OPIM are gaining attention in the recent years because of the importance of these processes in controlling polymer morphology and mechanical properties. In this method, the hot melt is injected to the mould. Later, a thin surface layer will be formed in the mould. Two pistons move reversibly at same frequency and the molten polymer is subjected to high pulse shear stress given by the reversely moving piston. The melt is allowed to cool down slowly. The layer shape was rectangle initially and gradually changes to an oval shape and then gradually transforms to round shape. The flow channel of the melt shrinks so that the shear rate gets stronger until the gate is cooled. Shear rate is not uniform throughout but it is lowest between the thin surface layer and skin layer while it is highest between the intermediate and core layer.

Researchers have examined the effects of processing variables on the mechanical properties of general-grade PP prepared by OPIM under low pressure. One of the researches was mainly concentrated on density as a function of oscillating holding pressure measured using density gradient columns. In this study, the existence of spherulites and shear-induced shish-kebab crystals was also confirmed by differential scanning calorimetry (DSC) measurements. As a result, the mechanical properties of PP were observed to be greatly improved by OPIM. (Young's modulus and tensile strength were increased from 1.4 to 3.0 GPa and from 31.0 to 57.8 MPa.)

Longitudinal (MD) and transverse (TD) mechanical properties of iPP self-reinforced composites (SRCs) prepared in a uniaxial oscillating stress field by OPIM indicated three types of mouldings (two biaxial and one uniaxial). Biaxial PP SRC had a 55–70% increase in tensile strength and more than four times improvement in impact strength in MD. In TD, it exhibited more than 40% increase in tensile strength and 30–40% higher impact strength. The difference in stress–strain behaviours of the OPIM mouldings in MD and TD was also noted.

The crystallinity and morphology of iPP SRCs produced by conventional injection moulding and OPIM studied by scanning electron microscope (SEM) and wide-angle X-ray diffraction analysis (WAXD) revealed that the OPIM mouldings exhibited three crystalline forms: α-, β-, and γ-phases. The WAXD results indicated that the outer shear region of the OPIM mouldings has the highest α-phase orientation and largest proportion of β-phase content.

High-density PE (HDPE)/ LDPE SRCs prepared by OPIM have improved toughness, tensile strength, and tensile modulus. Compared with HDPE SRCs, toughness of HDPE/LDPE composites is found to be improved while the tensile strength of them remained constant. SEM analysis of the multilayer structure of fracture surfaces of HDPE/LDPE specimens revealed that the central layer broke in a ductile manner, while the shear layer break was brittle. High orientation of macromolecules along the flow direction was the reason for the increased strength and modulus. DSC and WAXD revealed that co-crystallization had occurred between HDPE and LDPE.

5.2.7 Filament winding

A filament winding process was widely used for thermosetting plastics but recent developments support the processing of thermoplastic SRCs also. The filament winding process involves winding filaments, usually fibres or tapes over a rotating mandrel. Pipes, cylinders, and spheres are the major shapes produced by this method. Continuous reinforcements are wound over a mandrel until the desired thickness is achieved. Partially or completely automated systems are available for this process which reduces the labour and production cost (Figure 5.6).

The machine can be programmed for winding in helical or polar types. Helical type operates similar to the lathe. The carriage or the delivery eye moves back and forth releasing the fibres, while mandrel rotates continuously. Rotational speed of the mandrel and the speed of the delivery eye can be adjusted according to the requirement. This is called two-axes winding and used mostly for manufacturing pipes. Winding machines are available with four axes and six axes too depending upon the requirement for advanced applications.

A winding machine with additional cross-feed axis perpendicular to deliver eye travel and a rotating fibre payout head mounted onto the cross-feed axis are called

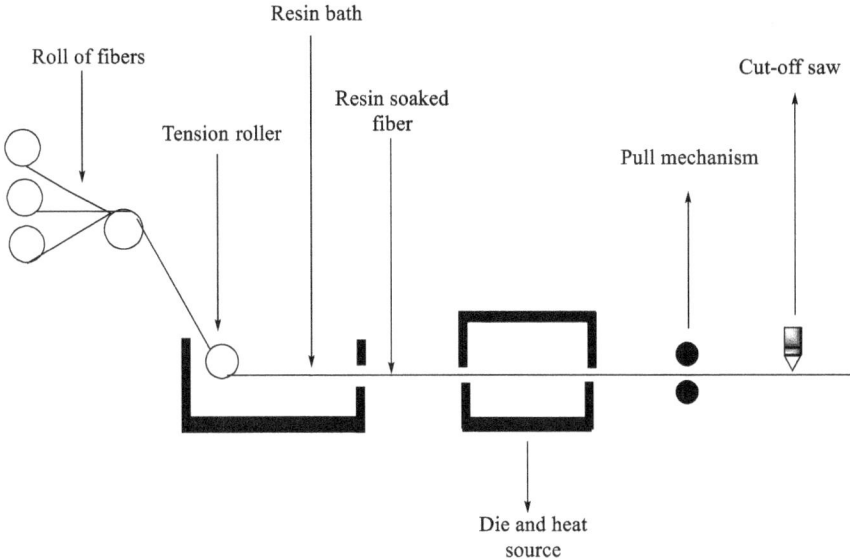

Figure 5.6: Filament winding.

four-axes winding machines. The payout head rotation can be effectively used for varying in width during winding. Six-axes winding machines have three linear and three rotation axes. Nowadays, most of the machines are computer or numerically controlled.

5.2.8 Laser-assisted tape winding

In this additive manufacturing process, a laser-assisted tape winding machine and a rotating mandrel are used for manufacturing the component by automated fibre and tape replacement (AFP/ATP). A computer or numerically controlled six-axes robot is used to install the AFP/ATP head, and the tape and substrate are heated with the laser followed by pressing the tape onto the substrate with the help of a roller in order to achieve better adhesion and consolidation [11].

5.2.9 Melt compounding

In this method, shear stresses by the viscous drag on aggregated fillers break them to the nanoscale aiding proper dispersion. Rupture mechanism and onion peeling mechanism were the two mechanisms proposed by the researchers in the dispersion of carbon black agglomerates (Figure 5.7).

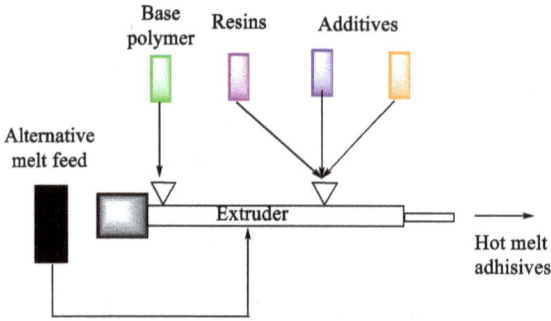

Figure 5.7: Melt compounding.

5.2.10 Impregnation

There are two types of impregnation processes for making SRCs: melt impregnation and powder impregnation.

5.2.10.1 Melt impregnation

In this process, reinforcement yarn is passed through a heated die into which the molten matrix polymer is injected in order to coat the yarn. The resulting rod is chopped into pellets (Figure 5.8).

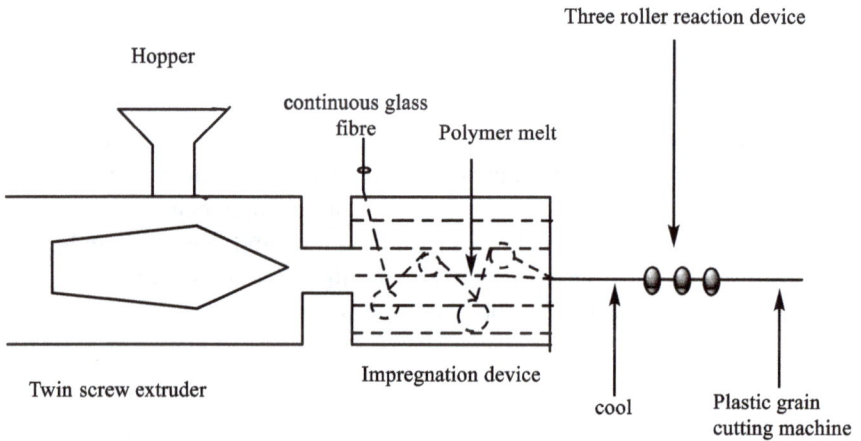

Figure 5.8: Melt impregnation.

5.2.11 Powder impregnation

This process is an impregnation process by using powdered materials. The sheets are passed through an electrostatically charged zone, which intimately distributes a fine powder into the substrate (Figure 5.9).

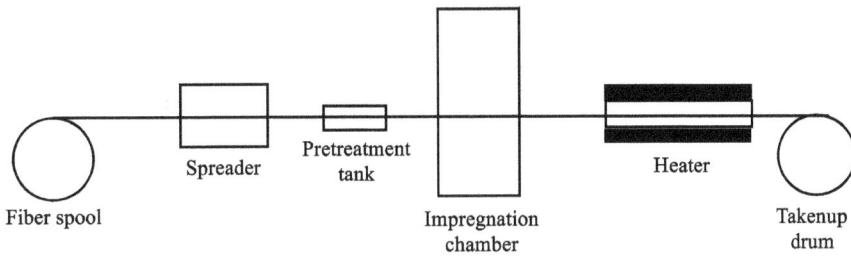

Figure 5.9: Powder impregnation.

There are a few other methods for the production of some particular type of SRCs. Tape winding and cold compaction followed by sintering, commingling, and so on are examples.

5.2.12 Additive manufacturing

Though most of the manufacturing techniques explained previously are additive in nature, researchers might be curious to know about the scope of conventional additive manufacturing techniques in the synthesis of SRCs. Popular additive manufacturing like fused deposition modelling, selective laser sintering, stereolithography, laminated object manufacturing, and laser-engineered net shaping are used for manufacturing composite materials. However, a notable work to develop SRPCs through any of these techniques is literally absent in the literature as the deposition of the material on itself as a fibre framework poses challenges in manufacturing composites with a good structural integrity.

References

[1]   C. Gao, L. Yu, H. Liu and L. Chen, "Development of self reinforced composites," *Prog. Polym. Sci.*, 767, 2012.
[2]   N. J. Capiati and R. S. Porter, "Concept of one polymer composites modelled with HDPE," *J. Mater. Sci.*, 10, 10, 1671, 1975.
[3]   J. Karger-Koscsis and S. Siengchin, "Single polymer composites-concepts, realisation and outlook," *Int. J. Adv. Sci. Technol.*, 7, 1, 1, 2014.

[4]   J. Karger-Koscsis and T. Barany, "SPCs: Status and future trends," *Compos. Sci. Technol.*, 92, 77, 2014.

[5]   A. Kmetty, T. Barany and J. Karger-Koscsis, "Self reinforced polymeric materials: A review," *Prog. Polym. Sci.*, 35, 10, 1288, 2010.

[6]   Y. Swolfs, Q. Zhang, J. Baets and I. Verpoest, "The influence of process parameters on the properties of hot compacted self-reinforced PP composites," *Compos. – A*, 65, 38, 2014.

[7]   D. Yao, R. Li and P. Nagarajan, "Single-polymer composites based on slowly crystallizing polymers," *Polym. Eng. Sci.*, 1223, 2006.

[8]   P. J. Hine and I. M. Ward, "Hot compaction of woven nylon 6,6 multifilaments," *J. Appl. Polym. Sci.*, 101, 2, 991, 2006.

[9]   P. J. Hine, R. H. Olley and I. M. Ward, "Use of interleaved films SRPC," *Compos. Sci. Technol.*, 68, 1413, 2008.

[10] H. X. Huang, "Continuous extrusion of SR high density PE," *Polym. Eng. Sci.*, 38, 5251–5253, 1998.

[11] S. M. Hosseini, I. Baran, M. Van Drongelen and R. Akkerman, "On the temperature evolution during continuous laser-assisted tape winding of multiple C/PEEK layers: The effect of roller deformation," *Int. J. Mater. Form.*, 2020.

Chapter 6
Characterization of self-reinforced polymer composites

Development of any new material requires complete understanding about the physical, chemical, and functional properties and structure of the material. Material science is a wide and deep area which requires integration of fundamental principles along with the tools developed by physicist and chemist. It is essential to study and correlate the material properties from various perspectives of basic scientific approach in order to recognize the lacuna in material research. Processability, manufacturability, and applications of self-reinforced composites (SRCs) were well examined with various characterization techniques so far.

There are different types of characterization techniques based on the property to be examined. Further, characterization tools may vary based on the fundamental principles used behind the operation of the tool. This chapter emphasizes on various characterization techniques used in polymer composite systems aiming at giving a brief idea of different characterization techniques to the reader. This will also be helpful for the reader to understand the upcoming topics easily. This chapter focuses only on some of the characterization techniques commonly used for self-reinforced polymer composites before and after mechanical testing so far in the literature. Major objectives of these characterization techniques are to study the physical, chemical, and microstructural changes in processing and testing, morphology, and fracture analysis. Interface characterization and changes in polymer properties like molecular structure and crystallinity after processing of SRCs are another interesting area to be studied. Macroscopic characterization like mechanical testing is discussed in a separate chapter due to vastness of the topic though thermal characterization is included in this chapter.

6.1 Microscopic characterization

Microscopic characterization is one of the oldest forms of material characterization. This is mainly used to study the composition and structure of the material. This is a very essential and important method because of the correlation between the microstructure and the properties of the material. Following are the various microscopic methods used by researchers based on the morphological features of the material to be analysed.

https://doi.org/10.1515/9783110647334-006

6.1.1 Optical microscopy

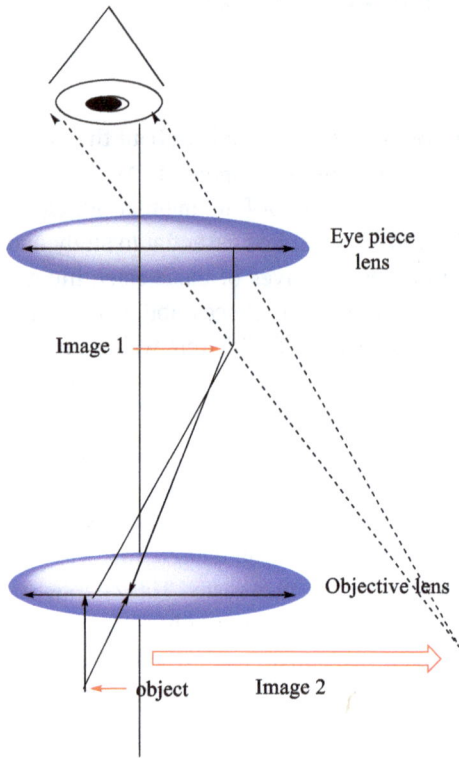

Figure 6.1: Optical microscopy.

Simple microscope and compound microscope are the basic types of optical micro-scopes. Simple microscope uses a single lens or group of lenses, while a compound lens uses a system of lenses in order to obtain larger magnification. In order to view the microstructure of polymers, proper specimen preparation is necessary. As an example, xylene etchant is used in the specimen preparation of polyethylene (PE). Specimen is immersed in the xylene solution for 1 min at 70 °C. Polished and etched surfaces will reflect more light back than that is reflected from the surface grooves. This difference will help in locating various surface properties (Figure 6.1).

Thermal stability of the oriented crystals was detected by the researchers using polarized optical microscopy. Presence of trans-crystalline layers is one of the features of SRCs, which can also be detected with the help of this method. Various other morphological studies are also conducted by using optical microscopic methods. Supermolecular structure of polypropylene (PP) SRCs is extensively studied by using optical microscopy. Quality of the laminates can also be examined and assessed by using optical microscopy after fabrication.

6.1.2 Scanning electron microscopy (SEM)

In scanning electron microscopy (SEM), a focused electron beam is used to scan the surface to be analysed (Figure 6.2). These electrons interact with the surface providing information about surface properties and microstructure. Very small sized samples can only be analysed by using SEM. Samples must be electrically conductive in order to get a proper image by SEM. Non-conducting samples like polymers will absorb charges. To avoid this they are to be coated with gold, gold/palladium alloy, platinum, chromium, tungsten, osmium, or graphite to make them conducting. Samples are to be dry as the specimen is subjected to electron beam in a vacuum chamber. SEM is widely used in characterization of SRCs.

Morphology and crystallinity of the material before and after processing, and nature of fracture and failure are analysed by using SEM coupled with fracture mechanics concepts. An overview of influence of thermal processing at various temperature ranges on fibre surfaces can be qualitatively examined using SEM. Fibre surface properties and their influence in interfacial properties can also be qualitatively examined.

In our earlier investigations, in order to study the failure pattern of interface tests in particular, we sputter coated the failed samples with a gold–palladium alloy and

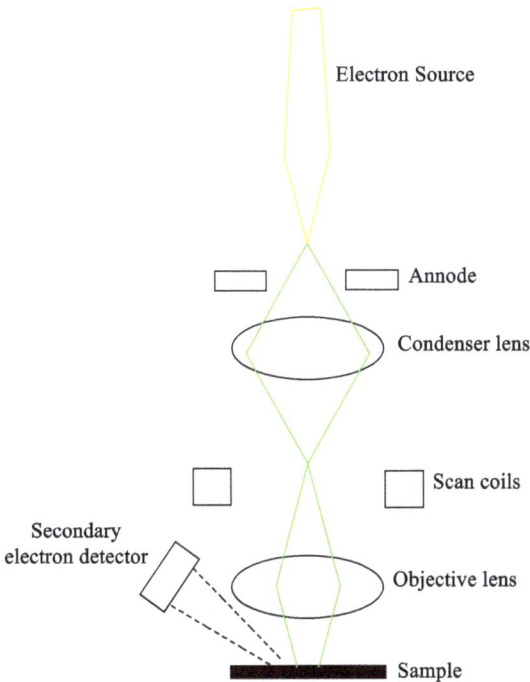

Figure 6.2: Scanning electron microscopy.

observed through the scanning electron microscope. ZEISS EVO 18 scanning microscope was used for the fractographic work with assistance from a smart SEM software. Electron beam generated by the lanthanum hexaboride (LaB_6) gun was made to be incident on the gold-coated samples. These non-conducting samples were earlier gold coated for 90 s. An optimum coating depth is required to get a clear image avoiding static electricity generation or "charge buildup." With a low accelerating voltage of 5 kV as the specimens are polymers, various fracture features could be studied from the micrographs. Extensive fractography was carried out on the failed samples of all the mesomechanical fibre pullout, quasi-static, and drop mass impact tests using Carl Zeiss optics of the SEM which could magnify the samples from 30× to 800× for clear texture or morphological studies.

6.1.3 Transmission electron microscopy (TEM)

In transmission electron microscopy (TEM), a beam of electron is passed through the ultra-thin (<0.1 µm) samples (Figure 6.3). The image obtained by the interaction of electron beam with the material is further processed and analysed to obtain the microstructural features. Image processing has been improved over the years since the development of the first TEM. In the analysis of cyanate ester self-reinforced composites, the structure of bisphenol A dicyanate analysed by using TEM revealed that the resin particles dispersed in the solvent were nanoscale and distributed at 40–60 nm [1].

6.2 X-ray spectroscopy and crystallographic methods

Electromagnetic spectrum of range of frequencies in light is effectively used in this method for characterization. Different range of spectrum can be utilized for this purpose. Atoms and molecules of the materials behave differently when some electromagnetic waves are passed through them. Microstructural features of the materials are explored by using some of the electromagnetic waves like X-ray, infrared (IR) waves, or even visible light.

Chemical and elemental properties of the material can be analysed by using X-ray spectroscopic methods. Atoms of the inner shell of the material are excited with a photon. When it regains its original state, the energy gets released as photon. Wavelength of this photon will be a characteristic feature of the element. Analysing the emission spectrum and by comparing it with the spectra of known composition yields qualitative and quantitative features of the material. X-ray diffraction techniques are widely used in characterizing SRCs.

Small-angle X-ray scattering/diffraction (SAXS) is a low-resolution crystallographic technique in which X-rays are passed through the samples, and the low-angle scattering nature of X-rays is recorded as a function of angle in the range of

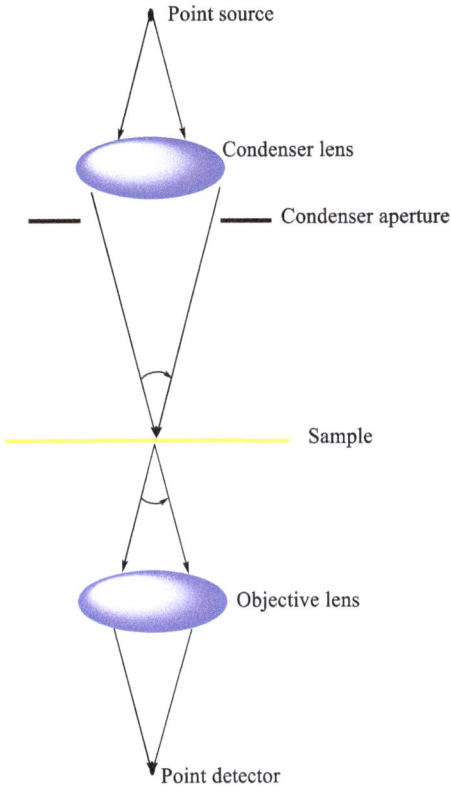

Figure 6.3: Transmission electron microscopy.

0.1–10 °C. The size, shape, and structure of colloidal size particles can be analysed by this method. The distance between the particles and pore sizes can also be analysed with this method. Researchers use this method to study the microstructure of biaxially oriented high-density PE SRCs along with wide-angle X-ray diffraction (WAXD), TEM, and DSC [2].

Wide-angle X-ray scattering/WAXD is one of the mostly used method in characterizing polymer SRCs and nanocomposites. WAXD analyses the Bragg peaks scattered to wide angle on much smaller length scale than SAXS. WAXD is widely used in assessing the degree of crystallinity [3]. Changes in crystallization fraction due to hot compaction can be analysed by WAXD. Molecular orientation is also studied by this method by some researchers. Shish-kebab structure is one of the popular crystalline structures of polymers, which can be visualized and analysed with this method. The α- and β-phase orientations and their locations for PPSRC are also studied by using WAXD [4]. Researchers identified the occurrence of co-crystallization with the help of WAXD and DSC [5].

6.3 Fourier-transform infrared spectroscopy (FT-IR)

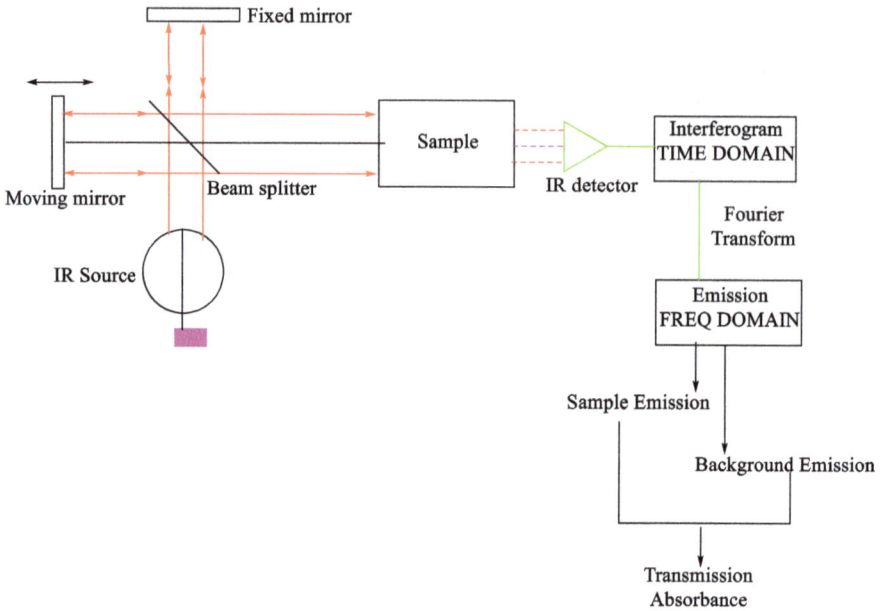

Figure 6.4: FT-IR.

Attenuated total reflectance–Fourier-transform IR (FTIR) spectroscopy is one of the most important polymer characterization methods in which an IR radiation beam is transmitted through the samples (Figure 6.4). Some amount of IR is absorbed, emitted, or transmitted through the thin specimen. The properties of materials are related to variations in the energy states of materials interacting with the radiation. Absorption causes changes in the vibrational energy states in the IR region. Based on the nature of absorption at a specific frequency level, information about the molecular structure can be obtained for the material. Authors carried out FTIR in IR affinity-1 machine on interfacial pullout test samples in order to analyse the possibility of any changes in molecular bonds when the interface is formed which support the interfacial properties. As there are no chemical changes expected while processing polymer SRCs, this characterization method is not popularly used by the researchers [6, 7]. However, in polymers exhibiting thermomechanical susceptibly due to forming and cooling, this is a good tool.

6.4 Thermal characterization

6.4.1 Differential scanning calorimetry (DSC) and thermogravimetric analysis (TGA)

Thermal characterization is to be carried out for the specimens kept either at a controlled heating rate or a particular temperature regime. The phase changes are to be studied by using this method, and the data may be utilized for processing SRCs and to analyse various properties of composites working under different temperature conditions. In order to determine various thermal events of each polymer component, and stability of these materials at processing temperatures, thermal analyses can be conducted. Due to their long molecular chains, polymers are materials with very complex behaviours. There are three major thermal events that can be observed in polymer materials: the glass transition, crystallization, and melting. DSC (differential scanning calorimetry) is a method for determining the thermal characteristics of materials. The principle of its operation is to measure the variation in the difference of temperature Δt between polymer sample and reference when temperature of the oven varies. This heat flow is directly proportional to the heat capacity of the material at a given temperature. TGA (thermogravimetric analysis) is a thermal analysis which consists of the measurement of the mass variation of a sample depending on time for a temperature or a given temperature profile. At different temperatures, chemical reactions can release gaseous species or form oxides resulting in a change of mass of the sample. This change in mass is registered according to the temperature.

DSC and TGA performed by the authors in SDT Q600 machine are shown in Figure 6.5. It helps in determining the exact processing temperature window. In our experiments, it was conducted on samples with a mass of about 7 mg extracted from the sheet, and the fabric at a rate of 20 °C/min with an N_2 purge of 100 mL/min up to 600 °C for sheet and 800 °C for fabric. Heat flow versus temperature plot of DSC curve explains all the events in thermal processing. TGA was performed for confirming the temperature of processing window in hot compaction. In the hot compaction method, matrix softens appreciably to compact, and the fibres remain intact. Thus, it has to be confirmed that the matrix softens and compacts, and the reinforcement is stable under the processing temperature. Similarly thermal analysis provides necessary data required for processing polymer SRCs. Most of the researchers have used DSC as a preliminary thermal analysis in their work.

6.4.2 Dynamic mechanical (thermal) analysis

This method is often used to characterize the viscoelastic polymer materials. Metals, ceramics, thermosets, and elastomers can also be characterized by this method. A dynamic load (oscillatory stress/strain) is applied to the sample with desired variations

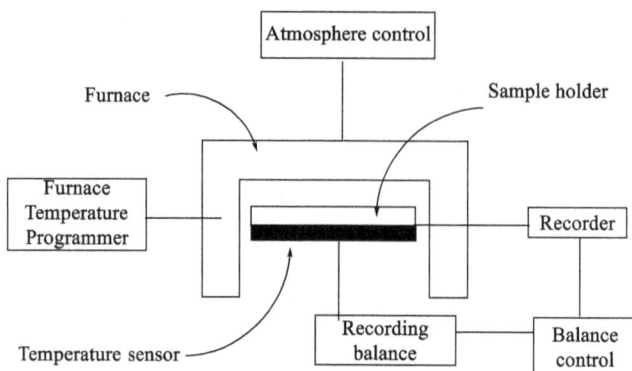

Figure 6.5: Differential scanning calorimetry.

in temperature and frequencies. The dynamic mechanical properties such as complex modulus (ratio of stress to strain) can be determined by this method. Complex modulus in pure viscoelastic materials, where the phase difference in stress and strain is $\pi/2$, is expressed as follows:

$$\text{Complex modulus, } E_C = E_S + E_L$$

where E_S is the Young's storage modulus and E_L is the loss modulus. Here storage portion represents the elastic nature and the loss modulus represents the viscous nature. Most of the viscoelastic materials generally exhibit properties in between that of ideal viscoelastic and ideal elastic materials with some phase difference. Mechanical performance of PPSRCs prepared by exploiting the polymorphism of PP was studied with different frequencies and temperatures by Abraham et al. [8]. Dynamic mechanical (thermal) analysis has been used to study PPSRCs made up of α- and β-polymorphs and with an ethylene copolymer by the researchers [3].

6.5 Acoustic methods

6.5.1 Through transmission C-scan

Composite structures are widely used in aircraft structures, which could be subjected to barely visible impact damages. This causes significant reduction in its structural properties which can be inferred through the C-scan technique which is a widely accepted method to locate internal damage in laminated composites. Matrix cracking, fibre failures, delamination, and so on can be analysed by this method.

Authors used an ultrasonic through transmission immersion C-scan as shown in Figure 6.6 to detect the internal defects in the post impact samples with a transducer frequency of 2.25 MHz, pulse voltage of 200 V, in water, and transducer focal

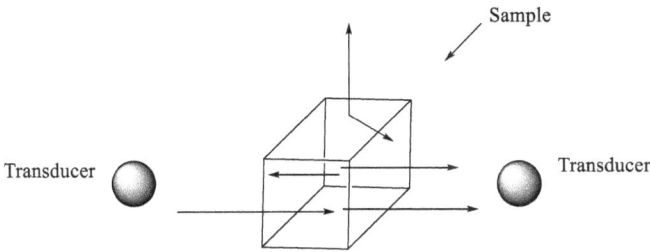

Figure 6.6: Transmission C-scan.

length of 2. The gain levels (35 db) were adjusted in such a way that the signal will not get saturated while scanning. C-scan should be coupled with other inspection methods to attain a proper conclusion about the failure nature. The samples of all systems were subjected to other types of analyses like visual inspection through light source and optical microscope, and cross-sectional analysis through the impact eye and SEM.

References

[1]   H. Cao, B. Liu, Y. Ye, Y. Liu and P. Li, "Study on the relationships between microscopic cross-linked network structure and properties of cyanate ester self-reinforced composites," *Polymers (Basel).*, 11, 6, 2019.

[2]   J. Lei, C. Jiang and K. Shen, "Biaxially self-reinforced high-density polyethylene prepared by dynamic packing injection molding. I. Processing parameters and mechanical properties," *J. Appl. Polym. Sci.*, 93, 4, 1584–1590, Aug 2004.

[3]   T. Bárány, A. Izer and J. Karger-Kocsis, "Impact resistance of all-polypropylene composites composed of alpha and beta modifications," *Polym. Test.*, 28, 2, 176–182, 2009.

[4]   L.-M. Chen and K. Shen, "Biaxial self-reinforcement of isotactic polypropylene prepared in uniaxial oscillating stress field by injection molding. I. Processing conditions and mechanical properties," *J. Appl. Polym. Sci.*, 78, 11, 1906–1910, Dec 2000.

[5]   G. Zhang, L. Jiang, K. Shen and Q. Guan, "Self-reinforcement of high-density polyethylene/ low-density polyethylene prepared by oscillating packing injection molding under low pressure," *J. Appl. Polym. Sci.*, 71, 5, 799–804, Jan 1999.

[6]   M. Sharan Chandran and K. Padmanabhan, "Microbond fibre bundle pullout technique to evaluate the interfacial adhesion of polyethylene and polypropylene self reinforced composites," *Appl. Adhes. Sci.*, 7, 1, 5, 2019.

[7]   M. S. Chandran, K. Padmanabhan, D. K. Dipin Raj and Y. Chebiyyam, "A comparative investigation of interfacial adhesion behaviour of polyamide based self-reinforced polymer composites by single fibre and multiple fibre pull-out tests," *J. Adhes. Sci. Technol.*, 34, 5, 511–530, Mar 2020.

[8]   T. N. Abraham, S. Siengchin and J. Karger-Kocsis, "Dynamic mechanical thermal analysis of all-PP composites based on β and α polymorphic forms," *J. Mater. Sci.*, 43, 10, 3697–3703, 2008.

Chapter 7
Interfacial mechanics of self-reinforced polymer composites

In fibre composite materials, matrix and reinforcement preserve their physical and chemical characteristics and together they exhibit mechanical properties which cannot be achieved if they are acting alone. These changes in properties are due to the interface between matrix and reinforcement. The interface is a layer that maintains the link between the matrix and reinforcement for the transfer of loads. There were attempts to characterize the mechanical, chemical, or physical properties of interfaces analytically. Atomic arrangement, molecular conformation, chemical constitution, morphological features of the fibre, or element diffusivity determine the kind of bonding in each system. Apart from hydrogen bonding, other adhesion mechanisms like adsorption and wetting, electrostatic attraction, chemical bonding, and exchange reaction bonding also play a role in the creation of interfaces. These mechanisms may exist alone or in combination at the interface [1, 2].

7.1 Micromechanics of stress transfer through the interface

In the interface perspective, the stress state and form of stress transmission provide a broader perspective on how fracture happens in composite materials. Micro- and meso-scale bond quality measurements at matrix fibre interface are carried out in the past for this purpose. Fibre augmentation test, fibre pull-out tests, and fibre push-out tests are the most common and sophisticated interface property evaluation tests existing so far. Micromechanical fibre matrix interface failures in these test methods are mostly studied with the help of shear-lag models in the initial stages of development and fracture mechanics approach. An elastic fibre-embedded matrix was exposed to uniaxial tension in the shear-lag model. This model was refined later by considering the matrix around the fibre possessing average features of the composite. This was one of the first attempts, and it laid the groundwork for future improvements in shear-lag models. There had been numerous other models developed in the same period to quantify the stress transfer through the interface with composite properties and by fibre fragmentation tests. Fracture mechanics approach provided improved models for interfacial shear stress (IFSS) in fibre fragmentation models as follows:

$$\tau_i(a,z) = \frac{a\sqrt{A_1}}{2}\left(\frac{A_2}{A_1}\sigma - \sigma_l\right)\frac{\sinh\sqrt{A_1}z}{\cosh\sqrt{A_1}(L-l)} \tag{7.1}$$

where $2L$ is the embedded fibre length, σ_l is the tensile stress at the fibre ends $z = L$, and l is the debond stress at the crack tip in the bonded and debonded boundaries,

https://doi.org/10.1515/9783110647334-007

A_1 and A_2 are complex functions of the constituents' elastic properties geometric factors [1, 2]. Energy-based fracture mechanical approaches are also developed extensively so far. Triangle- and trapezoidal-type traction–separation-based fracture mechanics computational models are widely used for interface modelling and simulation. This chapter focuses mainly on historical development of interface analysis in polymer composites and some of the recent developments in the interface characterization methods.

Multiple fibre pull-out models are other micromechanical models to evaluate interface properties developed in the 1990s in which a cylindrical fibre was assumed to be trapped inside a cylindrical shell of the matrix [3]. Later, finite element models were also developed based on stress- and energy-based fracture mechanics approaches like cohesive zone models. Though some self-reinforced composites (SRCs) possess excellent interface properties compared to conventional fibre-reinforced composites, micromechanics behind this feature is not much explored.

7.2 Mechanical test methods to evaluate interface properties

As the interface of matrix and reinforcement is invisible and nature of interface being highly unpredictable and least explored, the interface characterization and testing of interface properties require most sophisticated and accurate devices. Mechanical properties of fibre matrix interface were evaluated either by strength-based approach by assessing interfacial shear strength (IFSS) or energy-based approach through evaluating interface fracture toughness G_{ic}. Frictional strength is an important parameter as these tests are conducted in micro-composite systems. Popular interfacial tests can be listed as follows [1–3]:
- Single fibre compression test
- Fibre fragmentation test
- Fibre pull-out test
- Slice compression test
- Multiple fibre pull-out test

None of the tests has any standard method, and highly scattered data is predominant in all test methods. A micromechanical formulation to evaluate interfacial shear strength properties was developed by Chamis and Rosen [4]. A modification of fibre pull-out test is the base for this thesis, and a detailed literature survey which overviews important milestones leading to the work presented in this thesis is presented in this session in chronological order.

Classical works in microbond fibre pull-out test were reported by Bernard Miller and his colleagues in three consecutive articles since 1987. Single fibre microbond pull-out tests are presented in their first reported work. Pulling a fibre out of a matrix slab as performed in a traditional fibre pull-out may cause failure of the fibre before the interface failure. In order to address this issue, they made a microbond resin droplet around the fibre and by using a microvise the fibre was dragged out of the droplet. The force required to debond the fibre from the resin was recorded, and the required interface properties were evaluated. Glass, aramid, and carbon fibres (CFs) were tested in this manner [5].

In their second paper, they reported the influence of various testing parameters [6]. Initial tension in the fibre before the droplet shears was found retarding the shear strength. Loading rate was also found least influential. When the embedded length increases, shear strength was found to be decreasing, and in the interface, shear stress was not distributed evenly. It was also established that the predominant cause of failure was shear failure between matrix and the fibre by comparing the shear between the microvise–droplet and droplet–droplet. High scatter in the data was reported due to heterogeneity of fibre surface. Later, they studied the cause of the constant variations in bond strength results [7]. Inherent non-uniformity was confirmed to be the cause of bond strength variation. They also considered the effect of a compressive force near the vicinity of contact between microvise and the droplet and variations in contact angle. But these parameters were found to be least influential compared to fibre surface irregularity.

In a comparison of the fragmentation test with the microbond pull-out test in assessing the interfacial shear strength between CFs and a thermoset resin, Rao et al. [8] derived a relationship between the glass transition temperature of the matrix material and the blob size.. Smaller drops exhibited lower glass transition temperature, and they were not cured completely. With a modified curing environment and cycle, this issue could be resolved. The interfacial strength was still not matching with that of the fragmentation test results. Because of the use of diglycidyl ether of Bisphenol A (DGEBA) resin treated with a less volatile diamine curing agent, the results were found to be in good agreement with the fragmentation test findings.

Fracture mechanics-oriented energy release rate approach was proposed to reduce the issues in single fibre pull-out tests stated above [9]. The energy release rate of glass and CFs on a thermoplastic matrix was calculated using the finite element method. Influences of thermal stresses and frictional stresses were also analysed. Thermal stresses were found to be storing large amount of strain energy and resulted in increased energy release rates.

Piggott argued that the single fibre pull-out tests could be misleading due to the following reasons: (i) results are strongly influenced by the pressure and the frictional stresses which are not properly known parameters; (ii) mixed mode of failure in fracture; (iii) assumption of centro-symmetric debonding model is unrealistic; and (iv) least clarity on how the work of fracture is acting [10]. Thus, a modified method was essential to evaluate interface parameters.

Several initiatives to improve contact adhesion have lately been documented. The diameter of the fibre, the volume %, and the size of the micro-balloon particle all have a substantial impact on IFSS [11]. The IFSS evaluated using a microbond test of graphene oxide (GO)-grafted CF greatly improves the IFSS, according to Yu-nyun Ma et al. [12].

Zike Wang et al. discovered that a combination of electrochemical oxidation and sizing treatments reduced hygrothermal ageing related deterioration of inter-face adhesion in a hygrothermal ageing study on CF-reinforced polymer [13]. Ross F. Minty et al. [14] investigated the correlation between IFSS and matrix factors such as glass transition temperature, storage modulus, and linear coefficient of thermal expansion. In fibre fragmentation test, M. C. Seghini et al. discovered that the flax/epoxy system had the shortest critical fragmentation length and interfacial debonding length, as well as the highest IFSS values [15]. Junsong Fu et al. signifi-cantly improved the interfacial bond strength of CF-reinforced epoxy composites without lowering the fibre tensile strength using a layer-by-layer self-depositing technique of GO/silica multilayer films on the CF surface [16].

Samuel Requile et al. used a microdroplet debonding test to evaluate the hygro-mechanical behaviour of a single hemp fibre/epoxy bonding [17]. Lichun Ma and col-leagues improved the IFSS, interlaminar shear stress (ILSS), and dynamic mechanical properties of functionalized CF-reinforced composites by chemical grafting of octa-polyhedral oligometrix silsesquioxanes and their evolutionary amino-terminated hyperbranched structure on the CF surface [18]. According to Zijian Wu and col-leagues, better interface bonding of CF-reinforced unsaturated polyester with amino-functionalized carbon nanotube sizing agents increased ILSS and impact toughness. When CFs were oxidized, grafted, and then treated with silica nanoparticles, Guang-shun Wu and colleagues discovered that the ILSS of silanized CF composites increased marginally while that of hybrid (CF/Si) composites improved considerably [19].

Because of better adhesion between GO nanosheets and epoxy, the IFSS, ILSS, and flexural properties of GO/epoxy composites were considerably improved when they were chemically treated with cyanuric chloride and diethylenetriamine [20]. In another study on aramid fibres, Gong and colleagues discovered that dopamine-modified aramid fibres grafted with amino-functionalized GO reinforced in epoxy matrix enhanced the IFSS by 34% [21].

Understanding the interface characteristics aids in increasing the interfacial bonding of composite materials, which improves structural qualities. So far, many researchers have devised a variety of test methodologies [22, 23]. In comparison to other approaches, recent research demonstrates that the single fibre pull-out is one of the most reliable and straightforward ways for evaluating interfacial qualities. De-spite this, the single fibre pull-out test ignores the effect of fibre volume fraction as well as the need to statistically average test data. An embedded length shorter than the critical length of the fibre required for tensile failure should be maintained as the bonded contact between the matrix and reinforcement to avoid fibre failure before interface debonding. The multiple fibre pull-out tests eliminate these issues, albeit the slightly increased scatter of the pull-out load data must be considered. In the fu-ture, this technique can be further refined to reduce dispersion in the pull-out data.

Interfacial shear strength is a critical metric to determine in microbond pull-out experiments. This can be determined by taking into account the debond force at the time the first interface crack appears after external loading. There is a shear-lag period after this that cannot be reliably determined. Serge et al. proposed a novel method for estimating the IFSS that excludes the debonding force, claiming that it will increase the accuracy of interfacial shear.

A single fibre was pulled out of an epoxy matrix by Sastry et al. in a micro-bundle pull-out test [24] with limited success. The information gathered in each investigation was widely different. Hundreds of samples were evaluated to obtain a statistically significant average result for interfacial bond strength. Krishnan [25] created and tested the investigation given in this chapter for several composite systems. This new microbond fibre bundle pull-out methodology has been shown to be a more sophisticated and consistent test method for evaluating the interfacial characteristics because this method addresses the issue of volume fraction and is statistically averaged, considering the influence of defects in a cylindrical assemblage of fibres embedded in a matrix and their variance.

7.3 Meso-mechanical multiple fibre pull-out test

The approach described here is less time-consuming than some other approaches, such as Griffin et al.'s single fibre pull-out test from 1988 or Kelly and Tyson's fibre slab/resin pull-out method for interfacial examination from 1965 (Figure 7.1). Besides,

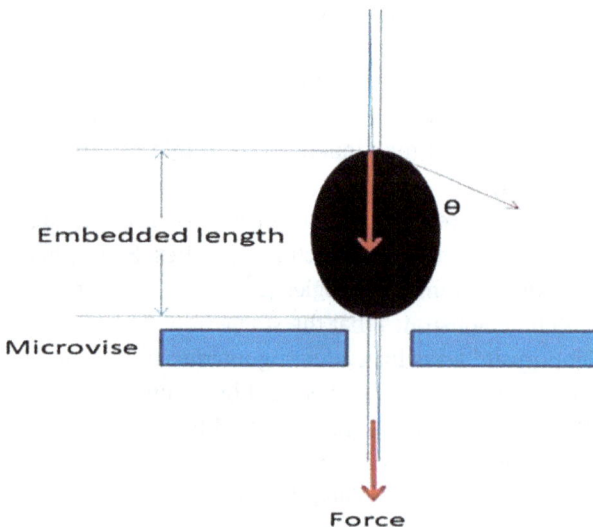

Figure 7.1: Schematic diagram of microbond bundle pull-out test.

single fibre pull-out tests are micromechanical in nature. The bundle pull-out test is more physically relevant, yields well-averaged findings, and overcomes the volume fraction issue [26].

A cylindrical assemblage model of fibres is considered in the analysis. The results may differ somewhat from the actual results due to the assumptions of cylindrical assemblage. Strength of the interfacial intrinsic bond (τ) is evaluated using the following equation:

$$\tau = \frac{F}{\pi \emptyset l} \tag{7.2}$$

where F is the peak debonding force minus any initial frictional forces, l is the length of the blob in the fibre bundle, and $\emptyset$ is the fibre bundle's average diameter. Microvise is a precision-engineered device that generates no frictional forces in the fibre or matrix. The area selected in the above computation is based on the assumption that failure occurs at the fibre matrix contact, which may be confirmed by keeping the drop size small. Higher drop size causes a larger matrix region to sustain the load and withstand interface failure, and the fibre may fail before the interface.

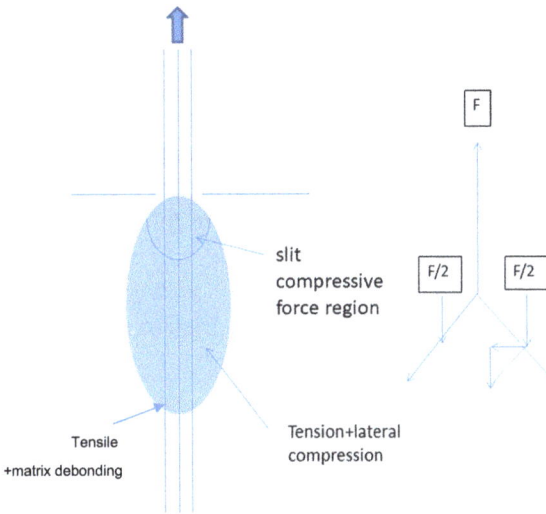

Figure 7.2: Schematic diagram of force resolution.

The ratio of tensile force to the shear force is also influenced by the embedded length. The diameter and the aspect ratio play important roles in the performance of the composite interface. The aspect ratio is the ratio of the fibre's embedded length to its diameter. If this ratio is lower than the critical value, failure occurs by shear. If the ratio is higher, failure occurs by tension. For polymer composites, this value is around 100 geometrically. Practically this may be higher than the geometric value too as the elastic modulus, tensile strength, and shear strength of the fibre with the matrix are also involved in addition to the aspect ratio. For a particular fibre diameter, if embedded length is lower, shear failure occurs. Thus, by increasing the diameter at a particular embedded length, we can assure a shear failure. Many parameters like size of matrix piece to be processed, temperature at which it is processed, viscosity of the matrix, time of processing, and how the samples are arranged for thermal processing are to be carefully optimized in order to get a proper blob of desired shape and length. If the underlying length is maintained below the critical point, multiple fibre pull-out test reduces the risk of fibre fracture before matrix shear. Again, when it comes to SRCs, the interface may be different compared to other systems because of the similar chemical composition of matrix and the reinforcement. So the practical aspect ratio may vary again from the theoretical values.

The angle of contact is a critical parameter in assessing interfacial properties, as shown in Figure 7.2. The matrix will be compressed against the fibre bundle by a force component acting perpendicular to the fibre bundle. The force is F Cot, where F is the applied force. The total of the interfacial debonding and frictional forces equals this force. The matrix gets compressed by the microvise in the vicinity of the point of application of the load (Figure 7.4). These locations are visible in micrographs too. In the other regions, the force will be longitudinal along with a lateral force since the Poisson's ratio of these materials is positive and it cannot be zero, meaning that there is no lateral force. Along with a longitudinal deformation, there exists a lateral force which will compress the fibre. This causes a component of force compressing the matrix to the fibre bundle at the appropriate place by acting in the lateral direction of the fibre. Total compressive force $F \cot\theta$ and expression for pressure can be obtained by this way. This can be calculated as follows:

$$F = F_{ab} + \mu F_c \tag{7.3}$$

High pull-out force (F_p) is the result of debonding force and the frictional forces and calculated by

$$F_p = \tau A + \mu F_f \cot \theta \tag{7.4}$$

where A is the interface area in mm^2, the static coefficient of friction (CoF) is μ, the bond strength is τ in MPa, θ is the contact angle of the fibre bundle with the resin droplet, F_f is the frictional force, and f is the frictional shear stress. Following debonding, the CoF varies from a starting value to a static value during frictional sliding. All of these steps can be seen on the load–deflection graph. IFSS is calculated using

$$F_p/A = IFSS = \tau + \tau_f \, \mu \cot \theta \tag{7.5}$$

CoF is manifested by the frictional tension, and the matrix shrinkage pressure will be the product of CoF and the pressure due to matrix shrinkage:

$$\tau_f = \mu(P_o + P_a = 1/2\ \pi r(dF_d/d_L) \tag{7.6}$$

P_o represents the matrix curing shrinkage pressure, and P_a represents the Poisson's contraction pressure, or stress relaxation pressure. These two variables have an impact on pull-out pressure as well. The bundle's radius is r, and the slope of the plot of maximum force versus embedded length in the section where embedded length exceeds the critical length is dFd/dL.

To determine the shear strength of the interface between the fibre and the matrix, a shear force is applied to displace and frictionally slide one phase relative to the other. This is accomplished by holding the matrix material and exerting a tensile force to the fibre. Small diameter fibres are expected to have poor breaking strength. The fibre breaks first if the force required to shear the interfacial bonding is greater than the fibre's ability to withstand tension, leading to inaccurate bond strength measurements. In the micro-bond pull-out process, the tiny interfacial contact area ensures that debonding happens before the fibre bundle ruptures and the matrix failure. Fabric failure prior to interface failure can also be eliminated using the multiple fibre pull-out technique.

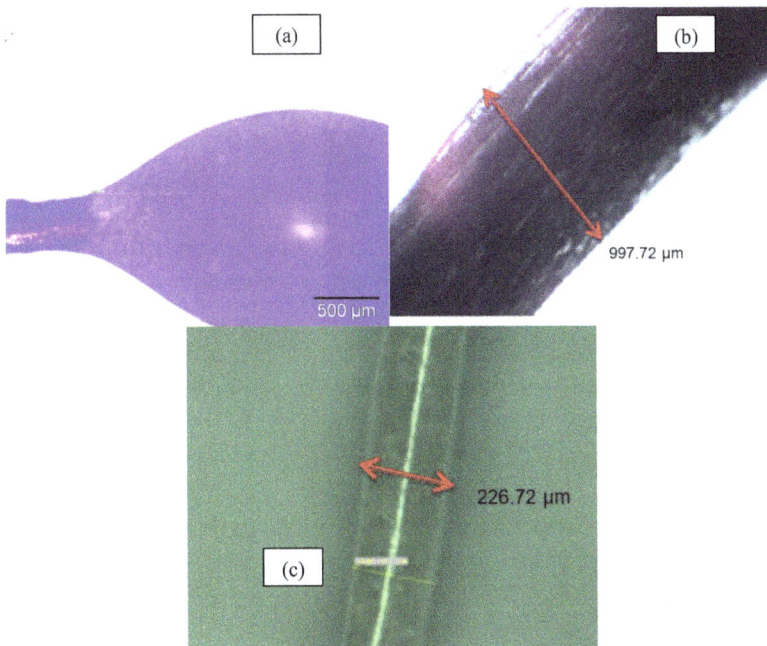

Figure 7.3: (a) Optical microscope, (b) blob under microscopy, and (c) fibre diameter measurement.

Fibre bundles taken from commercially available fabrics are used to make the samples. Optical microscopy is used to examine these fibres in order to establish their diameters (Figure 7.3). An Olympus BX 61-type optical microscope with reflection/ transmission and bright-field/dark-field operation capabilities is utilized for this purpose, with a maximum magnification of 1,000×. The removed fibre bundles are reintroduced into the hole created on micropolymer sheets after heat treatment, and the matrix forms a droplet or blob around the polymer fibre bundle, as shown in Figure 7.4. The fibres in the Instron were carefully placed and wrapped in aluminium foil to keep them together, with sand paper used to gain a good grip on them. This made it more likely that all of the fibre bundles would be taken out at the same time. After the matrix had hardened or cured, optical microscopy was used to quantify the fibre diameter and droplet size, and the interfacial contact area was estimated. Sprinkling of tiny fibres made up each fibre taken from the fabric. Each extracted fibre is referred to as a "fibre bundle" in this literature. Three of these "fibre bundles" are referred to as three fibre bundles, while one is referred to as one. The fibre bundles have an oval cross section. The fibre bundles, on the other hand, are thought to be cylindrical in shape.

Figure 7.4: Fabrication steps: (a) matrix pieces, (b) matrix–fibre bundle assembly, (c) thermal processing, and (d) final sample gets sheared of from the bundle on microvise.

Twenty-five samples were prepared for processing. After heat processing, 12–15 samples with satisfactory blob formation were chosen for testing. On the Instron 8801, a microvise was employed to grasp the droplet–fibre assembly (Figure 7.5). Under loading circumstances, a blob ripped off the fibre bundle at a strain rate of 2 mm/min. Maintaining a low strain rate minimizes compliance issues. In order to connect with the testing equipment, microvise is fastened with an aluminium cage. At one end of the fibre bundles, sand paper was used to hold the Instron, and the blob was supported by the slot in the microvise. While applying the displacement, the lower section of the blob is crushed by the microvise's slit, and the interface is sheared. Because the shear strength of the blob and the interface is less than the matrix's compressive strength, the blob shears through the fibre bundle. Because of improper or unequal blob growth surrounding the fibre bundle, some samples were discarded before being analysed. Several of the test findings were rejected when the individual fibre was pulled out instead of the entire fibre bundle.

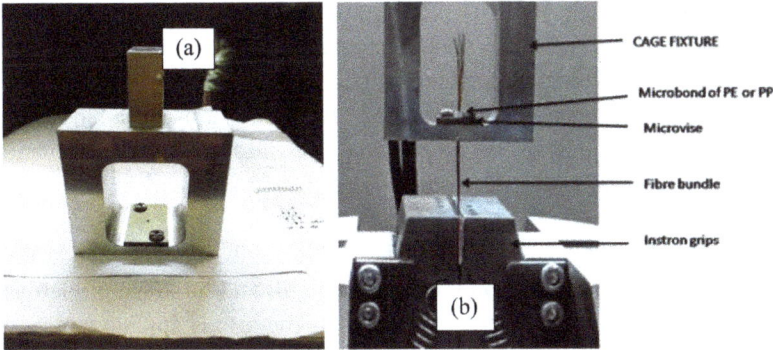

Figure 7.5: (a) Microvise and (b) testing.

A complete overview about SRC materials was obtained when we analysed these materials based on their chemical structure, design parameters, operating parameters, phenomenological concerns, and interfacial properties rather than analysing these materials just based on the mechanical properties alone.

Even if three systems of these systems of materials belong to the polyolefin class of materials, an additional CH_3 bond on polypropylene (PP) chain causes weak packing in the crystal structure of PP compared to more closely packed orthorhombic polyethylene (PE) chains. This could be one reason for weaker strength of PP SRCs under dynamic loading.

Thermal processing reduces the thermal stability of PP SRCs more when compared to PE SRCs which can be observed from the comparison of their thermogravimetric analysis data. PP SRCs form better interface with their own matrices compared to PE SRCs due to the lubricant nature of PE SRCs. Still, that could not help in its overall dynamic loading performance of these composites. PE SRC had very high improvement in overall mechanical properties (flexural modulus improved by 376%, Young's modulus by 100%, and % strain by 90%).

The multiple fibre bundle pull-out technique is a more reliable alternative to single fibre pull-out tests in many aspects since it is a statistically averaged test method that addresses volume fraction and decreases the likelihood of fibre breakage, and it is less reliant on blob length.

The single fibre pull-out test usually fails when the fibre load during tensile loading is less than the load taken by the fibre interface. Using the multiple fibre pull-out methodology, the interfacial shear strength of self-reinforced polymer composites was shown to be comparable to typical fibre-reinforced polymer composites, however, on the lower side. PP-based SRC showed higher interfacial strength (approximately 41 MPa) than other SRCs due to their great wettability. The PP matrix infuses into the spaces between individual fibres in microscopic analyses, resulting in a robust interface that was not observed in other SRCs, implying the explanation for the poor interface.

Ultra-high-molecular-weight PE (UHMWPE)/low-density PE (LDPE) SRCs showed interfacial strength of IFSS of around 5 MPa. Micrographs reveal that PP is more infused into the fibre bundle, and fractographic examinations back up the analyses' goal (reason for high IFSS). Weakest interface among all SRCs was that of Polyamide SRCs which reflected in reduction in its overall mechanical properties too. The nature of the fibre and matrix materials influences the method of interface analysis, highlighting the necessity for a good test method selection guideline to be developed.

Though PA SRCs having weak interface impact strength was commendable. The impact strength depends mostly on the fibre strength, and interface properties have least influence on low-velocity impact strength as the load is applied in a very short interval and in transverse direction. Energy of impact is utilized for fibrillation, brooming, and matrix damage in PP SRC. UHMWPE/LDPE composites are suitable for ballistic applications as it has very high impact strength and constrained damage area. An evolution coefficient between load, energy, size, and scale and time parameters was formulated which was found to be useful in analysing the evolution of performance of various materials under pull-out, quasi-static, and drop weight impact. It can be used as a quantitative tool to correlate the quasi-static properties with those of the dynamic ones. A high specific absorption value in low-velocity impact tests of PE and PP SRCs renders them to be useful in cargo and luggage applications in marine and aerospace domains at affordable costs.

References

[1]   J.-K. Kim, Y.-W. Mai and Y.-W. Mai, "Chapter 1 – Introduction," J.-K. Kim, Y.-W. Mai and Y.-W. B. T.-E. I. In: F. R. C. Mai, Eds. Oxford: Elsevier Science Ltd, 1–4, 1998.
[2]   J.-K. Kim, Y.-W. Mai and Y.-W. Mai, "Chapter 2 – Characterization of interfaces," J.-K. Kim, Y.-W. Mai and Y.-W. B. T.-E. I. In: F. R. C. Mai, Eds. Oxford: Elsevier Science Ltd, 5–41, 1998.
[3]   K. Padmanabhan, "Evaluation and methods of interfacial properties in fiber reinforced composites," Mechanical and Physical Testing of Biocomposites, Fibre-Reinforced Composites and Hybrid Composites (Elsevier), Chapter 18, 343–380, 2019.
[4]   E. J. Barbero, "Introduction to Composite Materials Design," Third Edition. CRC Press, 2017.
[5]   B. Miller, P. Muri and L. Rebenfeld, "A microbond method for determination of the shear strength of a fiber/resin interface," Compos. Sci. Technol., 28, 1, 17–32, 1987.
[6]   U. Gaur and B. Miller, "Microbond method for determination of the shear strength of a fiber/resin interface: Evaluation of experimental parameters," Compos. Sci. Technol., 34, 1, 35–51, 1989.
[7]   B. Miller, U. Gaur and D. E. Hirt, "Measurement and mechanical aspects of the microbond pull-out technique for obtaining fiber/resin interfacial shear strength," Compos. Sci. Technol., 42, 1–3, 207–219, 1991.
[8]   V. Rao, P. Herrera-franco, A. D. Ozzello and L. T. Drzal, "A direct comparison of the fragmentation test and the microbond pull-out test for determining the interfacial shear strength," J. Adhes., 34, 1–4, 65–77, Jun 1991.

[9]   C. Marotzke and L. Qiao, "Interfacial crack propagation arising in single-fiber pull-out tests," *Compos. Sci. Technol.*, 57, 8, 887–897, 1997.

[10] M. R. Piggott, "Why interface testing by single-fibre methods can be misleading," *Compos. Sci. Technol.*, 57, 8, 965–974, 1997.

[11] C. Zhi, H. Long and M. Miao, "Microbond testing and finite element simulation of fibre-microballoon-epoxy ternary composites," *Polym. Test.*, 65, December 2017, 450–458, 2018.

[12] Z. Wang, G. Xian and X. L. Zhao, "Effects of hydrothermal aging on carbon fibre/epoxy composites with different interfacial bonding strength," *Constr. Build. Mater.*, 161, 634–648, 2018.

[13] R. F. Minty, L. Yang and J. L. Thomason, "The influence of hardener-to-epoxy ratio on the interfacial strength in glass fibre reinforced epoxy composites," *Compos. Part A Appl. Sci. Manuf.*, 112, October 2017, 64–70, 2018.

[14] M. C. Seghini, F. Touchard, F. Sarasini, L. Chocinski-Arnault, D. Mellier and J. Tirillò, "Interfacial adhesion assessment in flax/epoxy and in flax/vinylester composites by single yarn fragmentation test: Correlation with micro-CT analysis," *Compos. Part A Appl. Sci. Manuf.*, 113, July, 66–75, 2018.

[15] J. Fu, et al., "Enhancing interfacial properties of carbon fibers reinforced epoxy composites via Layer-by-Layer self-assembly GO/SiO_2 multilayers films on carbon fibers surface," *Appl. Surf. Sci.*, 470, November 2018, 543–554, 2019.

[16] S. Réquilé, A. Le Duigou, A. Bourmaud and C. Baley, "Interfacial properties of hemp fiber/epoxy system measured by microdroplet test: Effect of relative humidity," *Compos. Sci. Technol.*, 181, June, 107694, 2019.

[17] L. Ma, Y. Zhu, M. Wang, X. Yang, G. Song and Y. Huang, "Enhancing interfacial strength of epoxy resin composites via evolving hyperbranched amino-terminated POSS on carbon fiber surface," *Compos. Sci. Technol.*, 170, November 2018, 148–156, 2019.

[18] Z. Wu, et al., "Interfacially reinforced unsaturated polyester carbon fiber composites with a vinyl ester-carbon nanotubes sizing agent," *Compos. Sci. Technol.*, 164, 195–203, Aug 2018.

[19] G. Wu, L. Chen and L. Liu, "Effects of silanization and silica enrichment of carbon fibers on interfacial properties of methylphenylsilicone resin composites," *Compos. Part A Appl. Sci. Manuf.*, 98, 159–165, 2017.

[20] L. Ma, et al., "Reinforcing carbon fiber epoxy composites with triazine derivatives functionalized graphene oxide modified sizing agent," *Compos. Part B Eng*, 176, July, 107078, 2019.

[21] X. Gong, et al., "Amino graphene oxide/dopamine modified aramid fibers: Preparation, epoxy nanocomposites and property analysis," *Polymer (Guildf).*, 168, 131–137, Apr 2019.

[22] S. Zhandarov and E. Mäder, "An alternative method of determining the local interfacial shear strength from force-displacement curves in the pull-out and microbond tests," *Int. J. Adhes. Adhes.*, 55, 37–42, 2014.

[23] S. Zhandarov and E. Mäder, "Determining the interfacial toughness from force-displacement curves in the pull-out and microbond tests using the alternative method," *Int. J. Adhes. Adhes.*, 65, 11–18, 2016.

[24] A. M. Sastry, S. L. Phoenix and P. Schwartz, "Analysis of interfacial failure in a composite microbundle pull-out experiment," *Compos. Sci. Technol.*, 48, 1–4,, 237–251, 1993.

[25] P. Krishnan, "A novel microbond bundle pullout technique to evaluate the interfacial properties of fibre-reinforced plastic composites," *Bull. Mater. Sci.*, 40, 4, 737–744, 2017.

[26] C. Y. Yue and K. Padmanabhan, "Interfacial studies on surface modified Kevlar fibre/epoxy matrix composites," *Compos. Part B Eng.*, 30, 2, 205–217, 1999.

[27] M. Sharan Chandran and K. Padmanabhan, "A Fractographic Study of PE, PP Self-reinforced Composites in Quasi-static Loading Conditions BT – Recent Advances in Mechanical Engineering." Springer Lecture Series, 603–618, 2020.

[28] M. S. Chandran and K. Padmanabhan, "Materials Today: Proceedings A novel correlative formulation of interfacial, quasi-static and dynamic behavior of polyamide self reinforced polymer composites," (In print), 2020.

[29] M. S. Chandran, K. Padmanabhan, D. K. Dipin Raj and Y. Chebiyyam, "A comparative investigation of interfacial adhesion behaviour of polyamide based self-reinforced polymer composites by single fibre and multiple fibre pull-out tests," *J. Adhes. Sci. Technol.*, 34, 5, 511–530, Mar 2020.

[30] M. Sharan Chandran and K. Padmanabhan, "Microbond fibre bundle pullout technique to evaluate the interfacial adhesion of polyethylene and polypropylene self reinforced composites," *Appl. Adhes. Sci.*, 7, 1, 5, 2019.

[31] M. Sharan Chandran, K. Padmanabhan, M. Zilliox and C. K. Tefouet, "Processing and mechanical characterization of self reinforced polymer composite systems," *Int. J. ChemTech Res.*, 6, 6 SPEC. ISS., 2014.

Chapter 8
Performance of self-reinforced polymer composites

An era of improved productivity and cost reduction through automation of industry has been gradually transforming into improved digitalization of industrial processes since the development of the idea of Industry 4.0. Minute details about the materials are essential for representing a material precisely in such digital platforms. Further, to use the composite materials to full advantage, their performance under different loading conditions should be properly understood. This will also help in avoiding unreasonably high factors of safety in design. Thus, sophisticated testing methods, standards, and specifications are vital to be developed to gather such material data. The design and performance of a product depend on their mechanical properties and working environment. Testing of materials is an integral part of research and development, product design, and manufacturing. There are standard test methods to evaluate the properties and assess the performance of these materials under actual service conditions. There exist abundant data in the literature related to the performance of a wide variety of self-reinforced polymer composites (SRPCs). In this chapter, we describe the mechanical testing of SRPCs for the evaluation of their mechanical properties and the temperature and environmental effects on their performance. Fracture mechanic aspects of self-reinforced composites (SRCs) are discussed in a separate chapter.

Determination of any properties of materials should be according to the guidelines of various bodies setting standards for testing. Test protocols in this chapter are adopted from the American Society for Testing and Materials (ASTM) and the most updated standards should be used to determine various properties of materials which will help incorporate the most recent developments in design.

8.1 Quasi-static properties

Tension, compression, shear, and flexural properties under quasi-static conditions are the basic need in the design of most of the composite structures. Testing procedures, the significance of these properties, and some of the examples in the literature are covered in this section.

8.1.1 Tensile properties

Test samples are subjected to a gradually increasing load in this type of test. A stress–strain diagram of the material is one of the essential data required to determine various properties of a component subjected to tensile load. Elastic and plastic

https://doi.org/10.1515/9783110647334-008

properties of a material under quasi-static loading can be predicted by using a load–deflection plot or stress–strain diagram obtained from a tension test. Tension test specimens can be prepared by injection moulding, compression moulding, hand lay-up, hot compaction, or any other methods. Sample shapes are usually flat dog bone shaped with rectangular cross sections for fibre-reinforced composites.

Materials that cannot be cut into dog bone shapes are tested with simple rectangular flat shapes. Rectangular tabs are bonded at the ends for better grip in such cases. Samples can be cut from a composite laminate by machining operations. However, processing methods and machining operations have a significant effect on the tensile properties of self-reinforced composites.

The tension tests of polymer composites are conducted following ASTM D7205/D7205M-21 standard in laboratory conditions of temperature and humidity, at a required rate of cross-head movement. Load and displacement data are recorded during the test. An extensometer or strain gauges can be used to measure the strain, and a stress–strain graph is plotted based on the output. The most commonly used test speed is 5 mm/min to evaluate tensile strength, Young's modulus, and so on. Some other properties like secant modulus, tensile strain, percentage elongation, yield strength, ultimate tensile stress and corresponding strain, and stress and strain at breaking load, toughness can also be obtained based on the requirement. The straining rate increases the tensile strength and modulus and reduces the elongation. Contact extensometers to measure elongation are being replaced by optical, digital, and laser-based devices these days. Stress–strain data of polymer composites also depend upon the time over which the load is applied and environmental conditions.

A valid tension failure is expected in the tension test. However, due to low shear strength or stress concentrations in the neck region of the dog bone shapes, undesirable failure may occur in samples. There are various methods adopted to avoid such situations. A schematic diagram of the standard specimen is shown in Figure 8.1.

Similar to all fibre-reinforced composites, for a fibre orientation of $\theta = 0°$, unidirectional SRPCs (fibres distributed along the direction of applied force) also experience a tensile failure of fibres followed by interface debonding longitudinally resulting in brooming of fibres in the failed cross section. This essentially leads to a linear stress–strain diagram up to the breaking point. For fibres oriented at 90° from the tensile load direction, the fracture may be due to matrix tensile/shear failure or interface failure. For off-axis samples (fibre alignment between 0° and 90°), the stress–strain relationship becomes nonlinear.

8.1.2 Compressive properties

Compressive properties evaluated from quasi-static loading are not a popular design criterion in most of the SRPCs even though there are many occasions in which

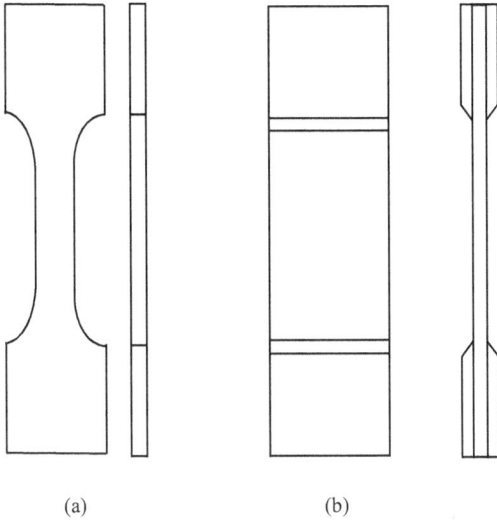

(a)   (b)

Figure 8.1: Specimens for tension test: (a) dog bone shape and (b) straight sided with tabs.

structures are subjected to compressive loads. This is because of the dependency on various other important factors on the compressive properties in SRPCs. Compressive loading may lead to linear or non-linear stress–strain behaviour in fibre-reinforced composite laminates. Though all the parameters associated with tensile loading are also associated with compressive loading, compressive strength and modulus are the parameters usually used in design processes. Thin samples used in the tension test would buckle under compressive loads. Providing side supports or employing block type of specimens are the alternate options. Transverse tensile stresses developed in the samples subjected to compressive loads may also lead to a brooming type of failure. The slenderness ratio is one of the different properties which differentiate the nature of failure between compressive failure and buckling.

To avoid buckling of thin samples during compression test, ASTM test standard ASTM D3410 is used with the test fixtures developed by Illinois Institute of Technology Research Institute (IITRI) (Figure 8.2). ASTM D6641 is the standard describing a recent method of compression test with combined axial and shear loads known as combined loading compression.

Compression after impact is a method of evaluating the residual compressive strength of the composite materials subjected to impact loading, mostly in low-velocity impacts. Invisible external damages due to impact events can lead to a reduction in the compressive strength of the materials. Samples for this test can be rectangular or circular in cross section and prepared by moulding or machining.

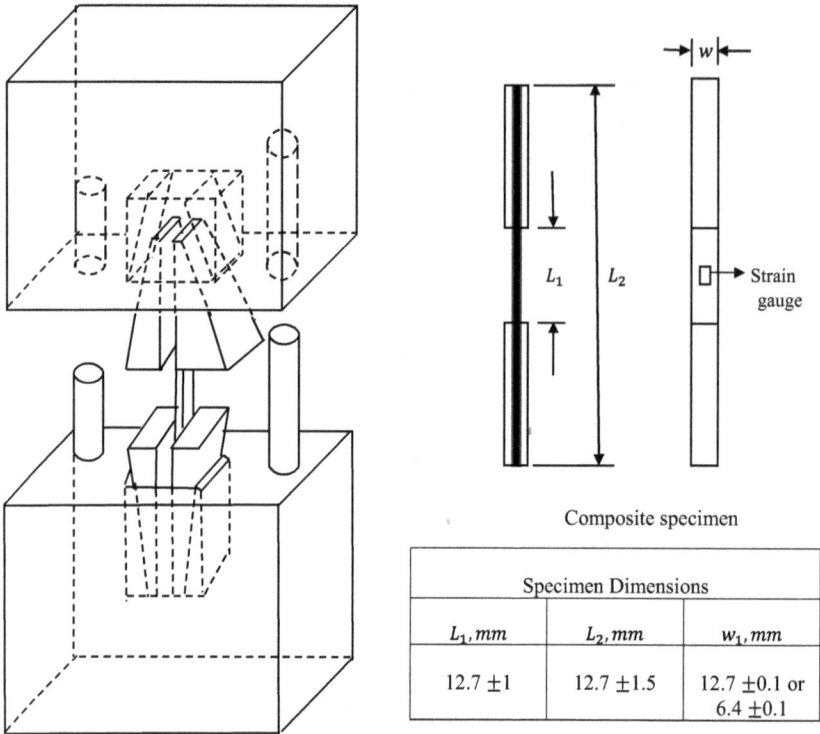

Composite specimen

Specimen Dimensions		
L_1, mm	L_2, mm	w_1, mm
12.7 ±1	12.7 ±1.5	12.7 ±0.1 or 6.4 ±0.1

Figure 8.2: IITRI compression test fixture.

8.1.3 Flexural properties

Flexural properties are very important in composite materials, especially in thin structures. Three-point bending (Figure 8.3) and four-point bending are the most common bending tests performed on these composites in accordance with ASTM D790 and D6272 standards for fibre-reinforced polymer composites, and bending stress is evaluated by using homogeneous bending theory. Another ASTM standard D7264 is also used for flexural tests in polymer composite materials. Three-point bending is usually for the materials that break at small deflections and four-point bending is suggested for the materials that do not fail under three-point bending and undergo comparatively larger deflections. Specimens are prepared with rectangular cross sections which can be either cut from laminates or moulded.

Flexural modulus, flexural rigidity, maximum bending stress, deflection, and so on are the major parameters evaluated from this test. In some cases, the contribution from shear stresses is also taken into account while calculating the deflection. The most important dimensional parameter in a flexural test is the span to thickness ratio

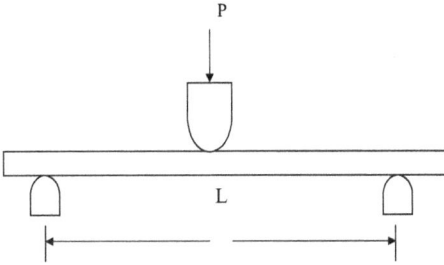

Figure 8.3: Three-point bending test.

(L/h). Various experiments conducted on SRPCs claim that the flexural strength is considerably high compared to the tensile strength. It is impossible to notice ultimate strength or any visible fracture in many SRPCs. Less rigid materials exhibit excellent resistance to failure but deform significantly. A rupture in the outer lamina can also be considered for extracting flexural properties. Progressive failure with debonding and delamination is the major failure in nature in SRPCs. Maximum stress developed under three-point bending on top or bottom lamina can be calculated from the following equation:

$$\sigma = \frac{3Wl}{2bd^2}$$

where σ is the bending stress (MPa), W is the load (N), l is the length of span (mm), b is the width of the cross section, and d is the thickness of cross section.

Flexural strength can be calculated using the above equation by considering the breaking load. When the maximum strain in the outer fibre reaches 5%, the corresponding flexural stress is considered as yield strength. In the cases of materials that do not break under the loading, the same criteria can be used to evaluate the maximum bending stress. Strain gauges attached to the surface of the samples can be used to measure the strains. Flexural modulus is very close to tensile modulus usually and can be evaluated from the slope of stress–strain plot. Fibre orientation in fibre-reinforced SRPCs plays a major role in flexural strength. Method of specimen preparation and machining operations also affect the flexural strength and modulus. These properties are inversely proportional to temperature. Modulus increases with higher strain rates, while the strength is significantly influenced by the span and cross-sectional dimensions. It is advisable to thoroughly investigate failed samples through fractographic methods to verify that the failure was caused by a pure bending stress. It is also essential to take proper care while testing in order to avoid undesirable effects like warping, twisting, or coupling effects during the test.

8.1.4 In-plane shear properties

There are many standard test methods available to determine the in-plane shear properties like shear strength and shear modulus. For valid testing, shear stresses are expected to develop in the plane of the laminate under test conditions, and an in-plane shear failure is anticipated. ASTM D3518 ±45° shear test, 10° off-axis test, and Iosipescu shear test are the popular in-plane shear tests conducted on SRPCs. There is a chance of tensile failure in these tests before the shear failure. Care should be taken to select the data from the samples which are subjected to an acceptable type of shear failure after the test.

Many methods are employed to evaluate the in-plane shear strength of the materials. A *picture frame test* method employs a thin square plate bonded at the edges, and the load is applied (Figure 8.4). In a *rail shear test* method (Figure 8.5), which is widely used for various tests in sandwich composite materials, shear properties can also be measured by using ASTM D4255 standard. However, the most popular and accurate methods for measuring in-plane shear stress are explained below.

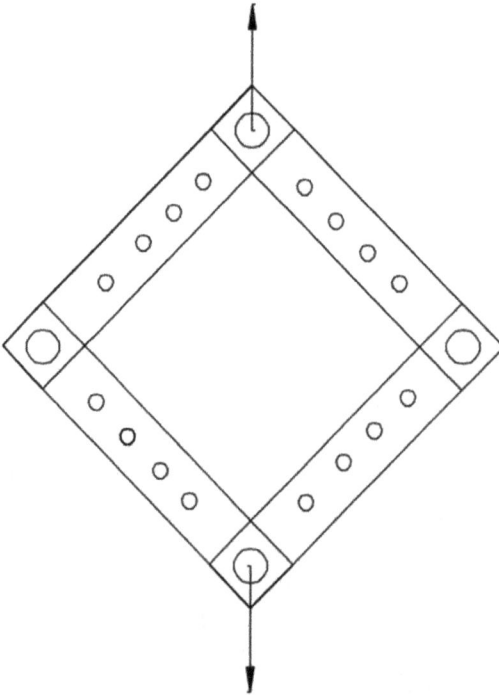

Figure 8.4: Picture frame test for in-plane shear test.

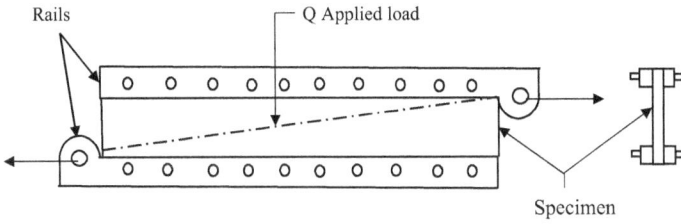

Figure 8.5: Rail shear test.

8.1.4.1 [±45]ₛ laminate shear test

In-plane shear stress developed in unidirectional composites is determined by a tension test conducted on a 1 [±45]ₛ laminate with an ASTM standard test procedure of D3518 proposed by Rosen (Figure 8.6). A combined load is developed on the lamina, and shear modulus can also be calculated from this method.

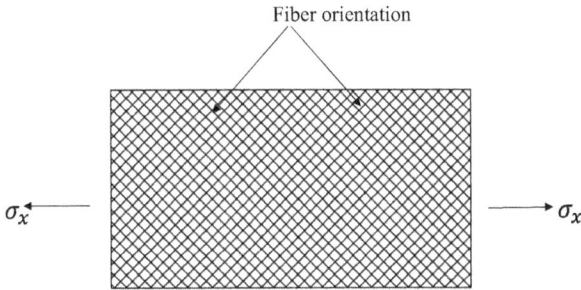

Figure 8.6: Loading pattern for tension test on [±45]ₛ coupon.

The concept behind this test method lies in the following relationships. Lamina shear stress and corresponding shear strain can be calculated from the laminate stresses and strains which are measured through the above-mentioned tension test. σ_x and τ_{xy} represent the laminate normal and shear stresses, and lamina stresses are represented by σ_{Lt} (longitudinal normal stress), σ_{Tr} (transverse normal stress), τ (shear stress), ε (normal strain), and y (shear strain). Lamina stresses and laminate stresses can be related as follows:

$$\sigma_{Lt} = \frac{1}{2}\left(\sigma_x + 2\tau_{xy}\right)$$

$$\sigma_{Tr} = \frac{1}{2}\left(\sigma_x - 2\tau_{xy}\right)$$

$$\tau = -\frac{1}{2}\sigma_x$$

Lamina strains can also be related to the laminate strains in a similar way:

$$\varepsilon_{Lt} = \frac{1}{2}\left(\varepsilon_x{}^0 + \varepsilon_y{}^0\right)$$

$$\varepsilon_{Tr} = \frac{1}{2}\left(\varepsilon_x{}^0 + \varepsilon_y{}^0\right)$$

$$\gamma = \varepsilon_y{}^0 - \varepsilon_x{}^0$$

8.1.4.2 Off-axis laminate test

In-plane shear properties of unidirectional composites can also be evaluated by an off-axis tension test as shown in Figure 8.7.

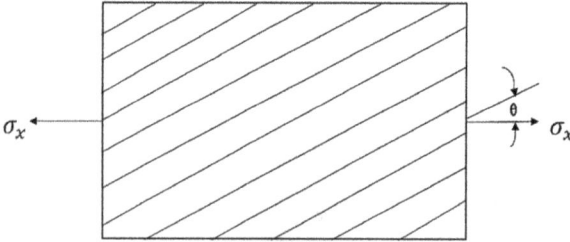

Figure 8.7: Tension test on off-axis coupon.

From the conventional strength of materials approach, longitudinal and transverse normal stress components are obtained as $\sigma_x \cos^2 \theta$ and $\sigma_x \sin^2 \theta$ and shear stress is $-\sigma_x \cos \theta \sin \theta$. This test will give good results when an off-axis angle of 45° is selected. At low off-axis angles, specimens should possess the high aspect ratio. Shear stress and shear strain along longitudinal and transverse directions of the lamina can be obtained from this test.

8.1.4.3 Iosipescu shear test

Publications of Adams and Walrath [18] explain details and recommendations of one of the most popular and easier shear test methods of composite materials called Iosipescu shear test. According to the ASTM D5379 standard, 51 mm length, 12.7 mm wide, and about 2.5 mm thickness flat rectangular specimens are used. Notches of 90° are cut up to the depth of 2.5 mm, on each side from the mid-length. This test will provide better result in cross-ply laminate, $[0/90]_s$. A schematic figure of the test loading mechanism is given in Figure 8.8.

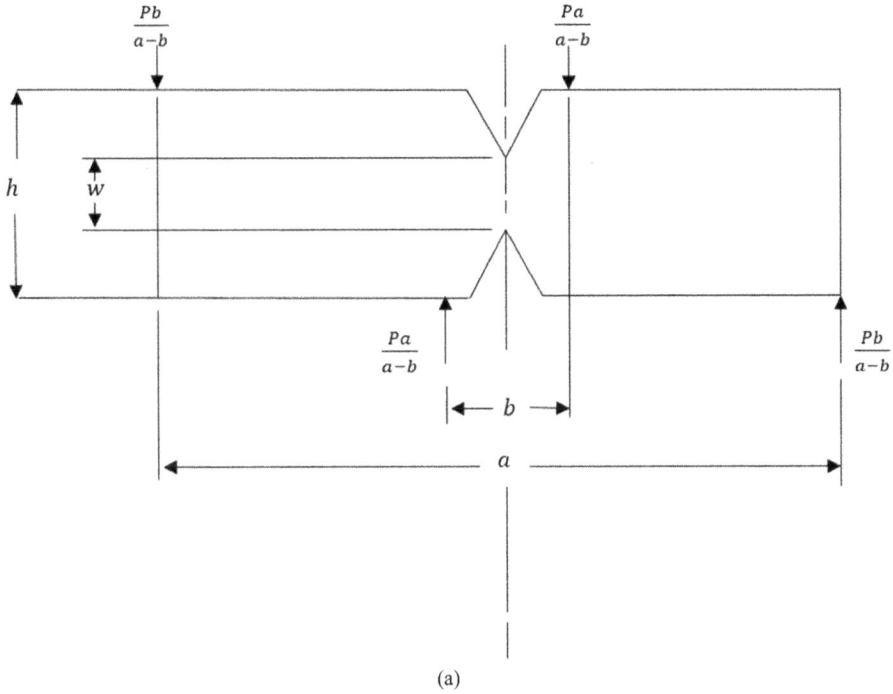

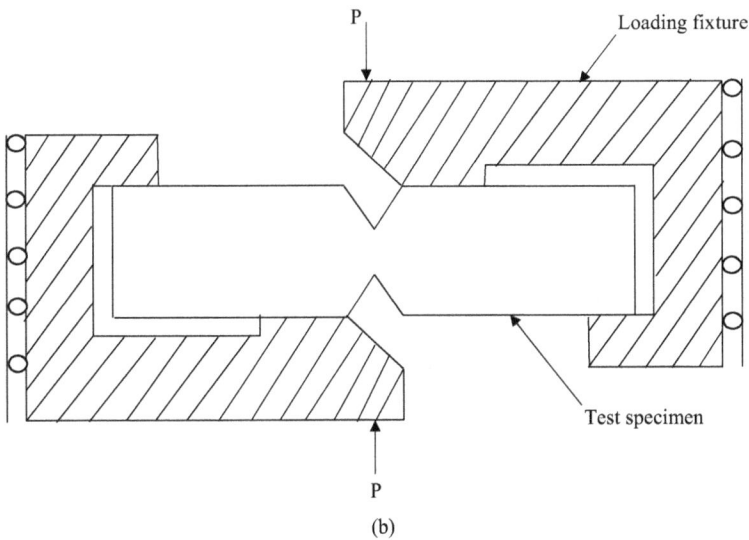

Figure 8.8: (a) Iosipescu shear test sample and (b) Iosipescu shear test loading mechanism.

When a compressive force is exerted on the fixture as in the figure, a state of pure shear without any bending moment is developed in the notched portion. This shear stress can be calculated as follows:

$$\tau = \frac{P}{bt}$$

where P is the applied force, b is the width between the notches, and t is the thickness (from cross section subjected to shear). The shear strength will be the load at which the specimen fails.

8.1.4.4 Torsion tube test

This test is conducted according to ASTM D5448 test standard (Figure 8.9). This method employs tubular specimens which make the sample preparation difficult. We are aware of the torsional shear stress and the corresponding strain developed when a twisting moment is applied on a circular shaft. The same principle is utilized in this test method. Pure shear stress can be induced in this method. Any kind of bending stresses or axial stresses are to be avoided while loading the sample. Precautions to avoid buckling of the sample are essential in this test. Strain gauges can be employed to measure the strains.

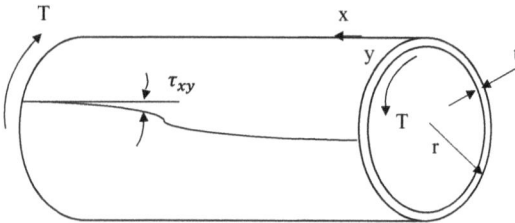

Figure 8.9: Torsion tube for shear test.

8.2 Fatigue properties

Since many of the structures of SRPCs are subjected to cyclic loading under service conditions, fatigue properties become an important parameter in the design and analysis of such structures. Components subjected to cyclic loads (fluctuating load, alternating load, or reversed loads) will fail at a stress value much lower than their static strength, and design of composite structures against fatigue loading is essential. Fatigue tests for SRPC components subjected to various cyclic loading conditions are discussed in this section. Most popular test methods are flexural fatigue test and tensile fatigue test. ASTM D7774-17 is the standard for flexural fatigue tests. Three-point bending and four-point bending are followed according to this standard. Fatigue resistance and the effect of processing, surface condition, and so on for large number of cycles can be evaluated through this test procedure.

Polymethyl methacrylate (PMMA) SRCs and polyolefin SRCs are studied for analysing fatigue properties so far. Three-point bending fatigue test was followed in an analysis of PMMA SRCs on Instron 1350 machine by Gilbert et al. [2]. The minimum-to-maximum stress ratio was set at 0.1 under a load control of 5 Hz, and maximum stresses ranging from 17 to 143 MPa. In each second, 250 load–displacement data were acquired, and the experiment was continued either until the failure of up to one million cycles. PMMA SRC was found to possess better fatigue strength and fatigue energy release rate compared to their pure form. SRPCs synthesized by UHMWPE fibres reinforced in ethylene–butene copolymer matrix developed by filament winding for biomedical applications were tested for fatigue properties by M. Kazanaci et al. [3]. ASTM D3479/3479M-19 is followed for fatigue loading with tension–tension cyclic loads. This study was conducted at room temperature with tension–tension cyclic loads with a loading rate of 8 N/s and minimum-to-maximum load ratio maintaining at 0.1 with a frequency of 1 Hz. The study was repeated on three different matrix proportions and different winding angles. More branched copolymers and wider winding angles were observed to be providing better fatigue strength in this study. Cyclic loads including compressive stresses are not common in SRCs due to buckling issues of thin structures. Most of these structures are not designed to support compressive loads. Thus, cyclic load involving pure compressive repeated stresses are not included in this discussion.

8.3 Impact properties

Energy dissipation and absorption capacity of materials under various impacts and shock–loading conditions is an important design criterion for many of the applications. The impact may be classified as low-velocity (<10 m/s), medium-velocity (10–50 m/s), or high-velocity impact (>50 m/s) based on the velocity of the object that hits the material.

Charpy and Izod tests can be used for impact tests of SRPCs as well (Figure 8.10). However, SRPCs are most susceptible to low-velocity impacts. ASTM D7136/D7136-M is used as the testing standard. The test can be performed by changing the impact energy levels either by changing the height or by changing the mass of the impactor. Critical impact energy may be obtained for samples made with different thicknesses. Perforation, penetration (partial penetration), and rebound are the three different possible impact events. The damage in low-velocity impact might be even due to a tool drop at production, service, maintenance, or repair site. Some of such unnoticed internal damages may get worsened under service conditions.

Low-velocity impact test results are highly dependent upon end supports, weight of the specimens and their dimensions, shape of the tup or striking part

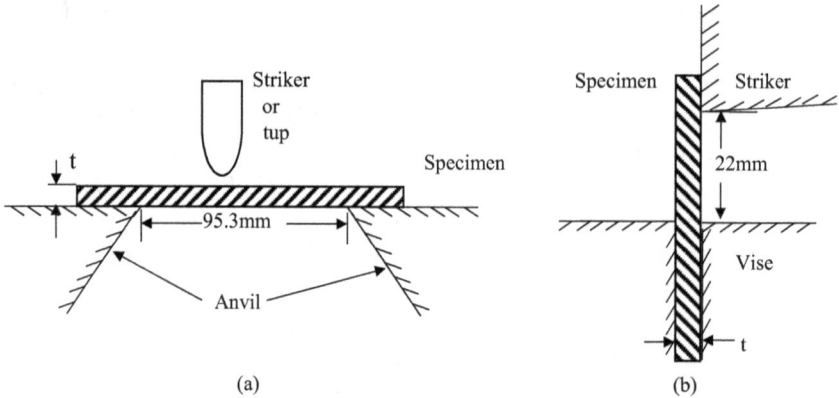

Figure 8.10: Impact test arrangements: (a) Charpy test and (b) Izod test.

along with the other parameters like fibre orientation, volume fraction, and processing parameters.

Impact studies determine the relationship between parameters like contact force, contact time, deflection, impact velocity impact energy, absorbed energy, frictional losses, and their relationship with damage mechanisms. Damage analysis methods like direct visual inspections, microscopic inspections, and ultrasound imaging are also used by researchers. This test method can be used for assessing the effect of fibre volume fraction, fibre orientation, order of stacking, processing parameters, and other things on impact resistance of the composite materials.

Test procedures for Charpy and Izod impact tests are similar to that of metallic samples. Specimen is kept in simply supported position in a Charpy test, and the pendulum strikes the specimen at the mid-span, while the specimen is kept as a cantilever beam in Izod test and the free end is struck with the pendulum. Notches are provided to concentrate the stresses at the desired locations of failure. Fracture energy or the energy required to propagate the fracture can be directly evaluated from the machine.

Some SRPCs like UHMWPE SRPCs can absorb very high impact energy. High-velocity impacts and ballistic impacts are useful for complete understanding of impact resistance of such materials.

8.4 Creep

Creep accounts for the time-dependent failure behaviour of the materials and measures the deformation of materials under a constant load as a function of time. Stress–strain-related information obtained from conventional testing methods is conducted over a short period. But the polymer composites are intended to use over a long service period. Polymer properties deteriorate when they are used over a long period. Hence, time-dependant property data is essential in designing a polymer composite. Creep and stress relaxation data are useful in designing composite structures to overcome the limitations of conventional stress–strain tests.

Tensile and flexural creep tests are the prominent creep testing methods. In tensile creep test, a constant load is applied to a tensile test specimen and the extension is measured as a function of time. Extension can be measured in different ways. One easy method is measuring a difference in previously marked gauge length in different time intervals and another method by employing the strain gauges. Flexural creep resistance measurements are carried out by applying a constant flexural load and measuring the deflections as a function of time. Deflections can be measured by a dial indicator gauge or electrical resistance gauges. Tensile creep curve and flexural creep curve are plotted from the measured parameters and interpreted to understand the creep behaviour of the materials.

8.5 Other properties

8.5.1 Short beam test

This is the easiest method to evaluate interlaminar shear stress. The method is similar to that of three-point bending test but the test is performed according to the ASTM standard D2344/D2344M-12. Span of the beam is kept relatively low in this test compared to that of the flexural test so that a high interlaminar shear stresses will develop, and normal stresses will be negligible. Span-to-thickness ratio is recommended to be four in this test.

8.5.2 Interlaminar shear strength tests

Interlaminar stresses are developed in the laminate due to a mismatch in Poisson's ratios and cross-coupling of adjacent plies. These stresses are considerably high at the interface and near the free edges. Interlaminar shear stresses may cause laminate failure and the strains cause matrix cracks on the free edge significantly during fatigue loading.

Interlaminar stresses can be normal (tensile) as well as shear in nature. Normal interlaminar stress that causes failure is often considered to be of the same value of transverse tensile strength of the lamina. Interlaminar shear stresses are also called thickness shear. Interlaminar stresses usually cause delamination in the composite laminates. In the following test methods, there are high chances for irrelevant type of failure of the samples. Such test results should be carefully eliminated for a better prediction of the interlaminar shear properties of the composite materials.

8.5.3 Notched plate test

Two notches are made on opposite sides of a flat laminate with rectangular cross section as shown in the figure. Interlaminar shear stresses will be developed in the shaded area between the notches. Depth of notches and spacing of notches should be properly decided to avoid tensile failure in the sample at the notch. ASTM standard D3846–08 provides details about this test Figure 8.11.

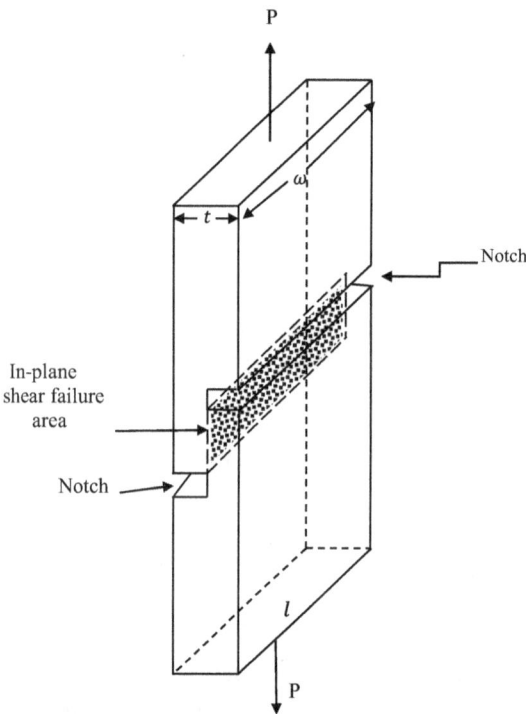

Figure 8.11: In-plane shear test.

8.5.4 Test method to evaluate fracture mechanics-related parameters

Fracture toughness is a very important material property which is defined as the resistance of the material against the propagation of flaws in the material. Strain energy release rate is the rate at which energy is released when fracture of a material occurs.

Stress state near the tip of a crack is predicted by using stress intensity factor. J-integral is also a measure of resistance of a material to crack propagation under a steady tensile deformation. Terms such as "critical" fracture toughness or "critical" strain energy release rate are often used if these parameters are critical values for the initiation of a crack.

An interlaminar fracture toughness test is conducted on a double cantilever beam according to ASTM standard D5528–13. Delamination of laminate composite materials is quantified with static interlaminar fracture toughness, and this test provides a load displacement from which critical strain energy release rate can be evaluated.

Samples for this test are straight cantilever beam type with rectangular cross section and 3 mm thick, 38 mm wide, and 229 mm long. While preparing samples for this test, number of layers should be even such that the layer thickness is same on both sides of the middle surface. An initial crack is required to be made during fabrication of the laminate by placing a 0.025 mm thick Teflon film. Portions of laminate on each side of the mid-surface deforms like a separate cantilever in opposite direction under loading. Load–displacement data up to 10 mm of crack extension are measured. Then the sample is unloaded and reloaded for each 10 mm extension in crack length up to a total crack length of 150 mm is reached. Details of the calculations of fracture mechanics parameters are explained in Chapter 9.

Strain energy release rate per unit width, $G = \left(P^2 C^2\right)/(bEI)$ and $(\delta/P) = \left(2C^3/(3EI)\right)$, where P is the applied load, C is the crack length, b is the width, E is the modulus of elasticity, I is the moment of inertia, and δ is the displacement of the edge due to the application of the load. Figures 8.12 and 8.13 illustrate various fracture mechanics tests.

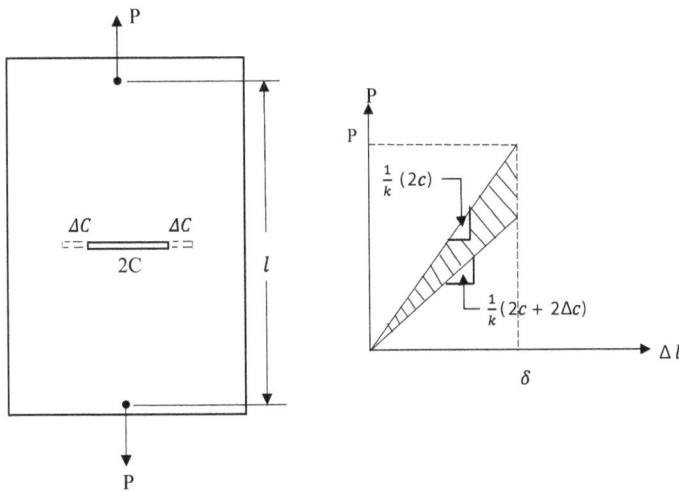

Figure 8.12: Tension test on a plate with crack to measure critical strain energy release rate.

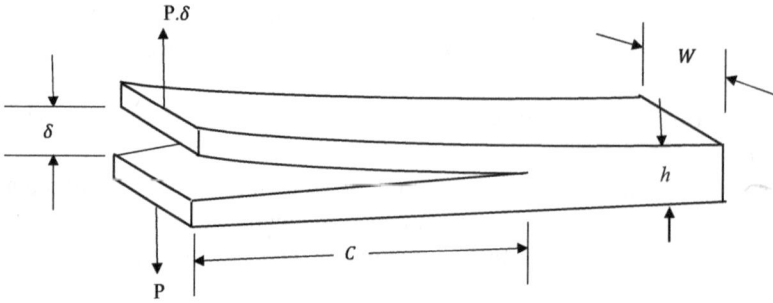

Figure 8.13: Double cantilever beam test for fracture toughness measurement.

Single-edge notched tension (SENT) and single-edge notched beam (SENB) are the tests those can be used to evaluate various fracture mechanics parameters. SENT test is a widely accepted in fitness-for-service assessments of flaws in pipeline girth welds under high strain conditions during installations and in subsea pipelines. SENB test is most appropriate in thermal cracking characterization. Fracture toughness value for the mode I fibre failure can be obtained from SENT tests. As the name indicates, specimens are made with single notch in this test and a displacement-controlled quasi-static load is applied. A tensile load is applied in SENT test and a transverse load is applied in SENB test. Both these tests are useful in investigating material behaviour over a range of temperature including ductile-to-brittle transition in metals.

8.6 Environmental effects

Various environmental factors like moisture, temperature, corrosive environments, or chemical interactions from surroundings may lead to degradation of performance of composite materials. This degradation could happen in various ways individually or simultaneously: (1) fibre degradation, (2) matrix degradation, or (3) fibre/matrix interface degradation and the degradation may be due to individual effects or combined effects of various environmental factors. As a result of these factors, strength, stiffness, or load-carrying capacity of composite structures will be compromised.

Studies show some of the following interesting features:

1. At low temperature or in non-corrosive environments, fracture strengths are relatively higher and independent of loading duration.
2. Corrosion rates are usually accelerated by the stresses developed during the loading.
3. Previous exposure to corrosive environments has less effect on properties.
4. Load-carrying capacity reduces significantly when any test is conducted with exposure to any of the previously stated environments.
5. Failure stress reduces with increase in temperature.

References

[1] B. D. Agarwal, L. J. Broutman and K. Chandrasekjara, "Analysis and Performance of Fibre Composites", 3rd Edition, John Wiley & Sons, 2006.

[2] J. L. Gilbert, D. S. Ney and E. P. Lautenschlager, "Self-reinforced composite poly(methyl methacrylate): Static and fatigue properties", *Biomaterials*, 16, 14, 1995, 1043–1055.

[3] M. K. Jacobs, D. Cohn and G. Marom, "Creep and wear behaviour of ethylene–butene copolymers reinforced by ultra-high molecular weight polyethylene fibres", *Wear*, 253, 5–6, 2002, 618–625.

[4] "Composite Materials: Testing and Design", ASTM STP 460, Philadelphia, PA: American Society for Testing and Materials, 1969.

[5] "Composite Materials: Testing and Design (Second Conference)," ASTM STP 497, Philadelphia, PA: American Society for Testing and Materials, 1972.

[6] "Composite Materials: Testing and Design (Third Conference),"ASTM STP 546, Philadelphia, PA: American Society for Testing and Materials, 1974.

[7] "Composite Materials: Testing and Design (Fourth Conference)," ASTM STP 617, Philadelphia, PA: American Society for Testing and Materials, 1977.

[8] "Composite Materials: Testing and Design (Fifth Conference)," ASTM STP 674, Philadelphia, PA: American Society for Testing and Materials, 1979.

[9] "Composite Materials: Testing and Design (Sixth Conference)," ASTM STP 787, Philadelphia, PA: American Society for Testing and Materials, 1982.

[10] "Composite Materials: Testing and Design (Seventh Conference)," ASTM STP 893, Philadelphia, PA: American Society for Testing and Materials, 1986.

[11] "13. Composite Materials: Testing and Design (Eighth Conference)," ASTM STP 972, Philadelphia, PA: American Society for Testing and Materials, 1988.

[12] "Composite Materials: Testing and Design (Ninth Volume)," ASTM STP 1059, Philadelphia, PA: American Society for Testing and Materials, 1990

[13] "Composite Materials: Testing and Design (Tenth Volume)," ASTM STP 1120, Philadelphia, PA: American Society for Testing and Materials, 1992.

[14] "Composite Materials: Testing and Design (Eleventh Volume)," ASTM STP 1206, Philadelphia, PA: American· Society for Testing and Materials, 1993.

[15] "Composite Materials: Testing and Design (Twelfth Volume)," ASTM STP 1274, Philadelphia, PA: American Society for Testing and Materials, 1996.

[16] "Composite Materials: Testing and Design (Thirteenth Volume)," ASTM STP 1242, Philadelphia, PA: American Society for Testing and Materials, 1997.

[17] "Composite Materials: Testing and Design (Fourteenth Volume)," ASTM STP 1436," Philadelphia, PA: American Society for Testing and Materials, 2003.

[18] "Composite Materials: Testing, Design, and Acceptance," ASTM STP 1416, Philadelphia, PA: American Society for Testing and Materials, 2002.

[19] G. L. Richards, T. P. Airhart and J. E. Ashton, "Off-axis tensile coupon testing", *J. Compos. Mater.*, 3, 3, 1969, 586–589.

[20] D. E. Walrath and D. F. Adams, "Analysis of the Stress State in an Iosipescu Shear Test Specimen," NASA°CR-l 76745.

Chapter 9
Experimental fracture and failure analysis in SRCs

Since Griffith's contribution during World War I, fracture mechanics approach has been evolving as the integral part of engineering design and analysis. It provides deep insight into the cause of various failures in engineering structures which would help in improving the designs by adopting proper measures to prevent the failure. Fracture mechanisms in self-reinforced composites presented in this chapter are based on the previously reported studies. This chapter will also provide an insight as to how fractography fits into the failure analysis of self-reinforced composite materials.

9.1 Introduction

Composite materials are anisotropic materials and inherent microscopic flaws are more common than their constituent materials. Heterogeneous behaviour of composite materials also makes them more susceptible to crack initiation. However, fundamental stress–strain relationships of composites materials are developed assuming a homogeneous flawless continuum material. Imperfections such as voids, delamination, and localized discontinuities act as failure initiation points in a composite material. The process of failure initiates with a microscopic flaw nucleation which leads to macroscopic crack. Then, the crack propagates and results in sudden fracture in very short period of time. This aspect of failure was neglected while developing the strength-based failure theories. Development and propagation of crack and fracture phenomena in self-reinforced composites are entirely different from metals or their parent polymers [1–3].

When a crack propagates, several models of failures can be identified. Tensile failure of nylon 6,6, PP and PE fibres observed through SEM are shown in Figure 9.1, 9.2 and 9.3. Fibre breakage can be seen near the crack tip. However, it has to be noted that these fibre breakages need not happen exactly in the plane of the crack. Fibres pulled out of the matrix and local debonding may occur in some composites near the crack tip as a result of crack propagation. There is another possibility of crack propagation without any fibre failure as well. If the fibre is more brittle and matrix is ductile, fibres may break even before crack tip approaches and a ductile failure of matrix occurs locally. A delamination is produced by a branched crack propagated through the interface [4–6].

Failure mechanisms are not common in all the SRPCs. There may be multiple mechanisms present in one system itself. In order to identify the failure mechanisms and prevent them, it is essential to know about these mechanisms and the tools that can be used. A few fractographic methods will be explained in the next session. Readers are advised to refer Chapter 6.

https://doi.org/10.1515/9783110647334-009

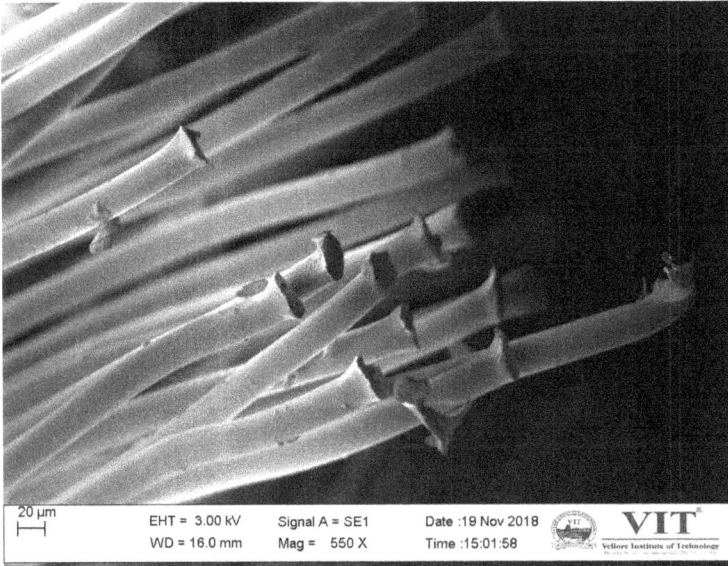

Figure 9.1: Tensile failure of nylon fibres.

9.2 Fractography

Fractography is the analysis of fracture surfaces which is helpful in identifying the failure modes and causes. Some applications of fractographic methods are listed:
- To find the cause of failure
- To locate the source of failure
- To identify the stresses responsible for crack initiation
- To understand the subsequent sequence of failure after crack initiation
- To study the material quality

In SRCs, fractography can be used for gathering information about the quality of the materials and manufacturing defects which would have resulted in crack initiation and consequent events. Evidence for such causes can be gathered by analysing the fractured surfaces. Micromechanics behind the development of damage and fracture can also be studied with the help of fractography. It can also help in linking the experimental study, numerical analysis, and validation. Fractography plays vital roles in predicting the failures, post failure analysis, and various case studies. However, there is not much attention paid in fractographic studies of polymer SRCs so far. There is a huge gap in this area and as a result of that there is a dearth in the literature available in this topic. In short, failure of SRCs can be classified into design failures, material defects, manufacturing defects, and in-service incongruities [7–10].

9.3 Methods and instruments

There are various steps to be followed carefully by an investigator in fractographic studies. It starts with a detailed collection of data and getting clarity about the aim and objective. Failed specimens are to be properly labelled before proceeding to any fractographic method. Next stage is to inspect the specimen meticulously without any tool, which is called visual inspection. Non-destructive test methods like ultrasonic scans or radiography can then be used for the next stage of inspection. Photographs of the specimen have to be preserved before proceeding to the next step for future reference. Specimen can be dissected further and macrographs of the failed surfaces can be taken. Detailed microscopic analysis can be carried out further with the help of various microscopic like optical microscopy or scanning electron microscopy [11–12].

Post failure damages result in vagueness in analysing proper cause of fracture and its development. Debris like broken fibres and surface crushing may hide some of the key features of the fracture such as riverlines and cusps leading to misinterpretation of the fracture details. Chemical reactions on the surfaces, contaminants on the fractured surfaces, and fracture surface anomalies caused by uncontrolled environments also cause issues in imaging. Proper care is essential while handling the specimen after the fracture.

Visual inspection can help in recognizing the fracture sequences, path of crack propagation, and the failure modes. Three modes of failures can be visualized in general

- Translaminar failure in the transverse direction of the laminate in which fibre failure occurs
- Intralaminar failure in the transverse direction of the laminate in which matrix or interface failure occurs
- Interlaminar failure in the laminate plane itself, in which the layers are getting separated.

Undamaged surfaces also have relevance in fractographic inspection. Any distortion on a surface indicates a delamination of internal damage. Crack initiation locations can be identified by nicks, dents, or splits. These evidences also suggest possible impact or wear damage. Orientation of gouges suggests the direction of an impact. Surface blistering indicates a possible delamination and a surface splitting suggests a ply splitting. These internal damages may also be due to buckling. Surface colour changes may be due to a chemical exposure or radiation. Like it was mentioned earlier, post failure damages can also indicate some of these features on the specimen [13–14].

If the specimen has undergone translaminar fracture, surfaces failed by tensile stresses are shiny or dark usually and flat and dull surfaces implicate a compressive failure. If the torsional stresses are responsible for the fracture, surfaces used have radial steps. Direction of rotation can also be predicted from the orientation of the steps. Chevron and radial marks indicate crack initiation and propagation directions.

Interlaminar and intralaminar fracture can even predict the modes of failures based on their surface features. Shiny fracture surfaces are often characterized by mode I interlaminar failure and dull fracture surface indicates a mode II interlaminar shear or surface fretting.

Though visual examination is useful in many fracture investigations, some of the fractured surfaces may be obscured by the presence of other layers or debris. It is to be noted that the internal cracks and damages contribute many important fracture events and visual inspection should be associated with non-destructive methods of assessment of failures to get complete idea of failure and fracture modes [15–16].

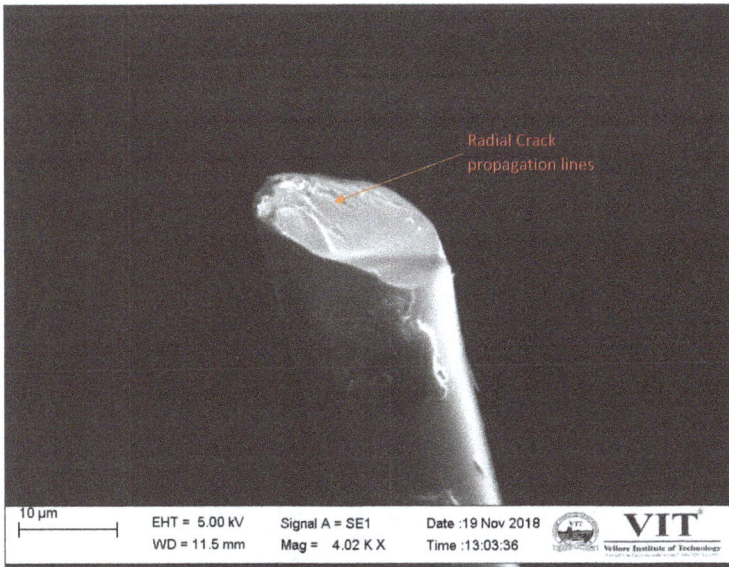

Figure 9.2: Tensile failure of PP fibres.

Visual inspection of fractured surfaces may not give complete idea about the failure. Hence, non-destructive methods of analysis are used for analysis of internal damages. Radiography methods, microscopic methods, spectroscopic methods, and optical methods are used for non-destructive internal damage assessment. Figure 9.4 is a SEM image of interface of PPSRC. These methods are coupled with visual inspection to gather better idea about failure modes, location of crack initiation and direction of crack propagation, relationship between internal and external failure modes, and the position of internal damage. Ultrasonic methods and dye-penetrant radiography are most relevant non-destructive method in fractography.

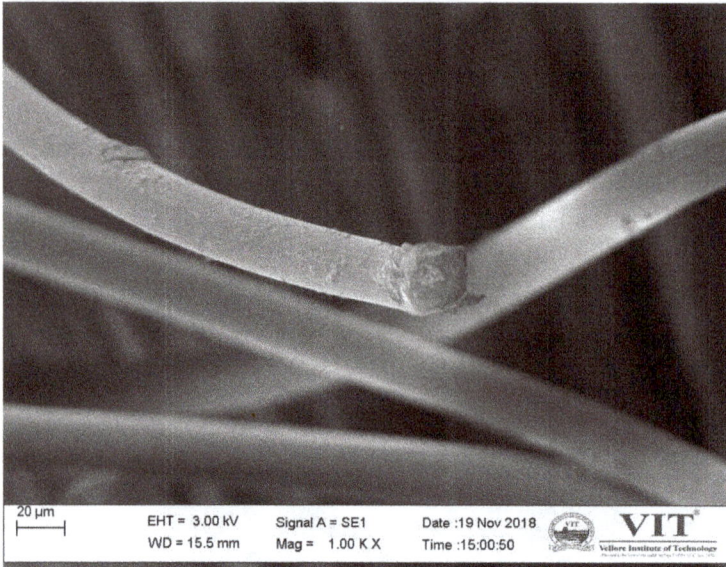

Figure 9.3: Cross section of PE fibre.

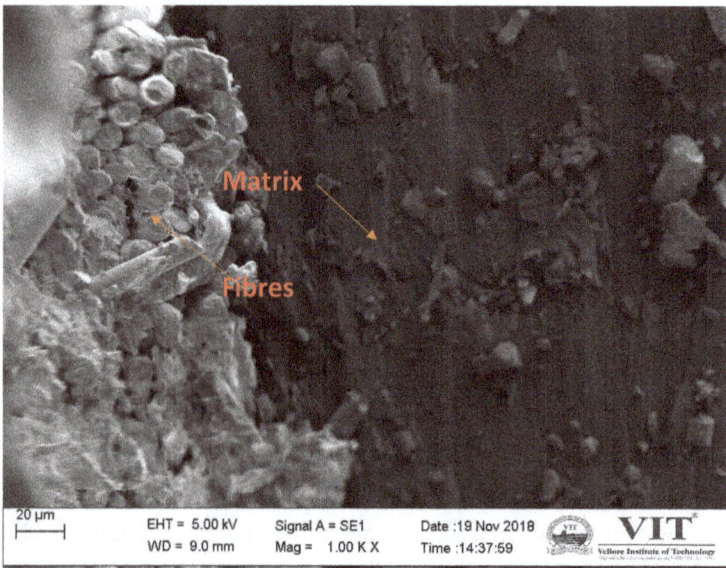

Figure 9.4: Fibre matrix interface region.

Fourier transform infrared spectroscopy is another tool which can be used to detect the source of failure through peaks that are uncommon in the material being analysed [17–19]. Thermomechanical analysis, which measures the dimensional changes in the material as a function of time and temperature, can also be used for failure analysis. Differential scanning calorimetry is a thermal method used in failure analysis. Details of these methods are given in Chapter 6.

9.4 Failure modes

Failure modes are generally identified as fibre-centred failure, delamination failures, fatigue failures, failure due to defects, or in-service failure. Features of these failures are separately discussed in this section. Failure by static loading and dynamic loading are different and fatigue failure is discussed in separate section.

9.4.1 Fibre-centred failures

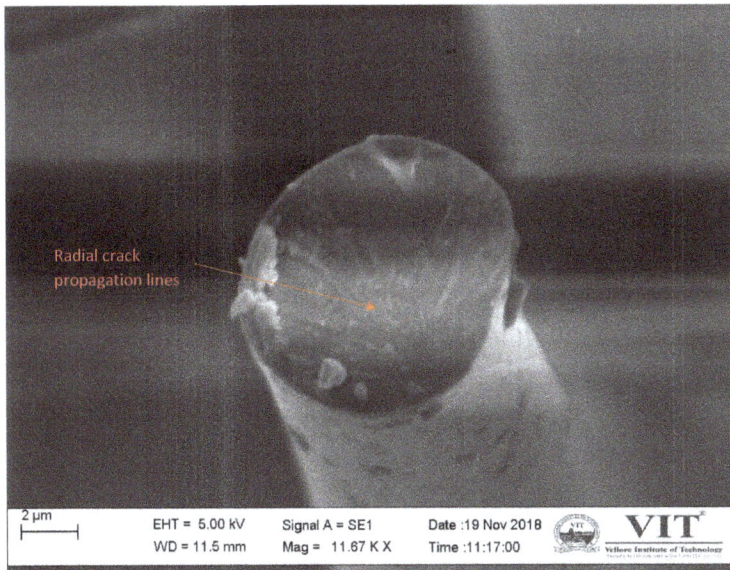

Figure 9.5: Polypropylene fibre cross section with crack propagation lines.

Figure 9.5 shows an example of fibre centred failure in PPSRC. Fibre ruptures are associated with translaminar failure. Most of the composite materials are designed to withstand either tensile or compressive failure under static loading. Pure tensile or compressive failure can happen under flexural loading for some composites. A combination of tensile and compressive failure can also happen under flexural loading. In-plane translaminar failure is also discussed in this section.

In slow fracture of unidirectional brittle fibres of self-reinforced polymer composites, failure initiates at tiny defects and radially propagates to the other regions. All fibres in the section will fail in the same plane with flat surface. But if the loading rate is increased, after the crack propagates from the initiation zone with increased speed, the plane of the fracture diverges resulting in rough topography and radial lines of fibres. These features can be identified in macroscopy. Under microscopic inspection of these failed fibres, radials starting from the defects on failed surfaces can be visualized clearly. Surfaces near the initiation used to be flat, gradually turning into rugged and rough as the distance from the initiation location increases. These radial patterns provide information about the local crack growth patterns. It can also be observed that failure initiated at one single fibre at the defect and failure of that fibre induced failure of other fibres too. This is called directly attributable fibre failures. If the fibres are not in direct contact, local failure started at a fibre propagates through the matrix to the next fibre. By tracking these events of local fracture initiation and propagation, global laminate failure features, and crack growth directions can be deduced.

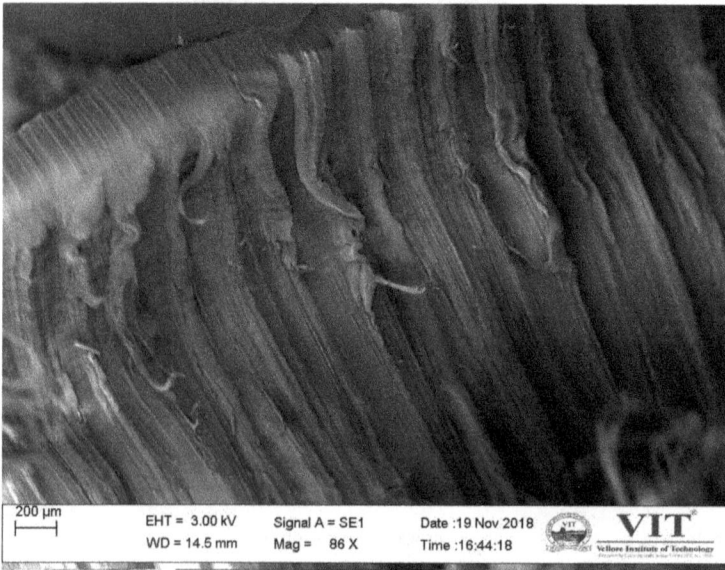

Figure 9.6: Polyamide matrix surface with impression made by the fibres after delamination.

Figure 9.5 shows a typical brittle fracture of fibre in tension. Fractured fibre surface has a central void and radial line emanating from that source of crack can be identified from figure 9.5. Surface is flat around the void and becomes rough at the outer surface.

Fibre brooming is an interesting feature in tension failure. A longitudinal ply splitting before the fibre breakage is the peculiarity of this feature. When a tension crack grows, there will be an associated shear stress developed across the crack. When this crack reaches near the fibre, crack divert to a direction parallel to the fibres resulting a shear stress in the fibre/matrix interface. This causes localized interface failure and debonding. But this type of crack growth results in blunting the crack and prevents further crack growth. If the strength of the interface is high, debonding through crack propagation will be limited resulting in a flat and smooth brittle fracture of fibres. Degree of brooming in failure is an important tool in interface failure analysis.

In multidirectional laminates, degree of brooming is a tool to understand the effect of delamination on the failure mechanism. Brittle fracture is accompanied by the failure of all the plies in the same plane. In such cases, it is observed that the secondary damages like matrix cracking (ply splitting) or delamination are limited. In cross-ply laminates, in which 0° plies are the load bearing layers, failure is initiated at 90° layers as matrix cracking and subsequent delamination of 90° layers with adjacent plies.

Compression failures are difficult to analyse because of their masked surface by debris. Microbuckled fibres are a peculiar feature of surfaces subjected to compression failure. PEEK is more prone to microbuckling. Compressive failure pattern is often associated with other types of damages such as longitudinal splits and delamination. Formation of 'kinkbands' used to happen in compression failure is a clear indication of compression failure. Compression failure of multidirectional laminates are a combination of different failure modes. Zero degree plies are more susceptible to compression failure with microbuckling while the off-axis plies often fail by i-plane shear and final fracture is characterized by delamination and matrix cracking. As compressive loads can also induce flexural loads, delamination due to flexural loading has been observed in such failures. Degree of delamination, whether it is local or global, has significant influence in the failure mode. In-plane compression failure causes an angled crack mostly in the middle of the laminate. Thus, degree of delamination and in-plane fracture are useful in differentiating the flexure and in-plane compression on the laminate. As mentioned before, it is quite impossible to determine crack growth direction in unidirectional laminates. However, it is possible to determine the crack growth propagation direction in multidirectional laminates.

Flexural failure of self-reinforced composites has features of fibrous tensile region and flat compression zones with middle distinguishing neutral axis. Similar to the features explained in tension and compression failures, radial, riverlines, and DAFF are associated with flexural failure regions and kinkbands and microbuckling are associated with compression failure under flexural loading in unidirectional and multidirectional self-reinforced polymer composites. In-plane shear failure is also caused due to microbuckling of fibres.

9.4.2 Delamination centred failure

Figure 9.6 represents delamination failure in PPSRC. Delamination may develop all of a sudden during loading or due to several past events which might have occurred in the service period. Delamination is a predominant feature of failure in compression failure. A composite structure may be subjected to compression either by direct loading or as a bending load. Interlaminar stresses are the prime reason for delamination in composite structures. Delaminated layers will be exposed to moisture and contamination which could lead to further failure of the structures. Delamination of most of the structures is a combination of different modes of failure such as mode I, mode II, and mode III. Matrix cracking and fibre/matrix interface failure can be observed through fractography in delamination failures.

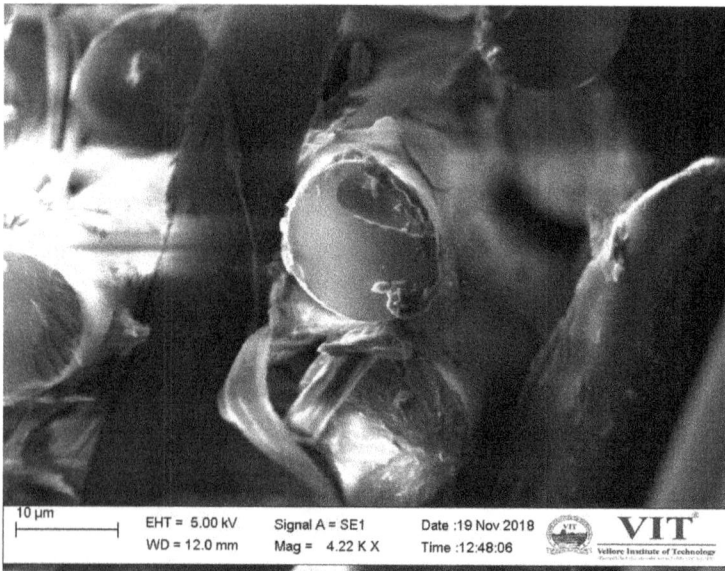

Figure 9.7: Porosity in polypropylene fibre.

One of the important features in self-reinforced polymer composite fractures is the formation of crazes under excessive tensile loads resulting in microvoid formation in a plane normal to the stress. Microscopic voids in the polymers are stabilized by plastic deformation and the adjacent fibres prevent further development of the craze. If the matrix is brittle, a failure type called 'cleavage' can be observed. As the energy of fracture is low and lack of plastic deformation in brittle materials result in these types of failures. Feathering and textured microflow are another type of structural failure observed in self-reinforced polymer composites. These are appeared with granular structure with clumps in the fracture propagation direction.

However, in thermoplastic self-reinforced polymer composites, textured microflow is rarely reported. Scraps, ribbons and riverlines are the matrix failure patterns observed in self-reinforced polymer composites. Crack initiates at the fibres and gradually develops in the matrix. Figure 9.7 shows porosity in polypropylene fibres and Figure 9.8 is the cross section of a PP SRC laminate.

Figure 9.8: Cross section of a PP SRC laminate.

References

[1]   ASM international, "ASM handbook – volume 11 failure analysis and prevention." ASM Handbooks Online, 2002.
[2]   J. Scheirs, "Compositional and Failure Analysis of Polymers." Wiley, 2000.
[3]   P. Stumpff, "Composites – failure analysis. ASM handbooks online," ASM International, 2003.
[4]   R. A. Grove and B. W. Smith, "Compendium of Post-Failure Analysis Techniques for Composite Materials." United States, 430, 1987.
[5]   ASM international, ASM Handbook – Volume 12 Fractography. 2002.
[6]   A. C. Roulin-moloney, "Fractography and Failure Mechanisms of Polymers and Composites." Elsevier, 1989.
[7]   D. Hull, "Fractography: Observing, Measuring, and Interpreting Fracture Surface Topography." Cambridge University Press, 1999.
[8]   S. J. W. Baas, "GARTEUR AG14: Fractography of composites." GARTEUR TP No 083. 1994.
[9]   R. J. Kar, Composite Failure Analysis Handbook. Volume 2. Technical Handbook/ Part 2. Atlas of Fractographs. 500, 1992. United States.

[10] K. Armstrong, W. Cole and G. Bevan, "Care and repair of advanced composites." SAE, 2005.

[11] B. Robinson, Health and Safety Issues Regarding Composite Failure Analysis. 4-1-2008.

[12] R. J. Parrington, "Fractography of metals and plastics," *J. Fail. Anal. Prev.*, 2, 16–19, 2002.

[13] A. Lemascon, P. Castaing and H. Mallard, "Failure investigation of polymer and composite material structures in the mechanical engineering industry." International Metallographic Conference. MC95, 10–12 May 1995. Materials Characterization 36 (4–5), 309–319, 1996. Colmar, France, Elsevier. Mater. Charact. (USA).

[14] D. Purslow, "Composites fractography without an SEM-the failure analysis of a CFRP I-beam," *Composites*, 15, 43–48, 1984.

[15] R. J. Kar, "A study of current and new fractographic techniques required for composite failure investigations." ASM Int. ISTFA 1988: International Symposium for Testing and Failure Analysis. The Failure Analysis Forum for Microelectronics and Advanced Materials. Conference Proceedings, 31 Oct–4 Nov 1988. 321–331. 1987. Los Angeles, CA, USA.

[16] R. A. Smith, "Advanced NDT of composites in the UK," *Mater. Eval.* 65, 7, 697–710, 2007.

[17] F. L. Matthews and R. D. Rawlings, "Composite Materials: Engineering and Science." Chapman & Hall, 1994.

[18] M. Hiley, GARTEUR AG20 Fractographic Aspects of Fatigue Failure in Composite Materials. DERA/MSS/MSMA2/TR000168/GARTEURFinalReportTP112. 2001.

[19] S. M. Freeman, "Characterization of lamina and interlaminar damage in graphite/ epoxy composites by the deply technique." Composite Materials: Testing and Design (6th Conference). 50–62. 1982. Phoenix, AZ, USA, ASTM, Philadelphia, Pa, USA. ASTM Special Technical Publication.

Chapter 10
Self-reinforced polymer composites – global solutions and technological challenges!

The search for natural and bio-based plastics has become a necessity due the decrease in the fossil resources and environmental and social concerns due to the strong greenhouse signatures of petroleum-based products. The quantum of bio-derived resins and bio-derived plastics that can be produced from the conventional, organic, and genetically modified (GM) plants is immense because India and the south-east Asian nations are world class producers of sugar cane, sugar beet, and other tubers such as vegetables with starch, cashew, badam, castor oil, and soya bean. These bio-products are rich resources of thermoplastic and thermoset polymers when derived through synthetic or other processes. This chapter attempts to project the actual possibilities of the bio-resin and bio-plastic market in this region and provides the knowhow for their production and utilization. Among all the other sources, cashew nut shell liquid (CNSL) is an abundantly available natural source for synthesizing phenolic compounds and high-temperature resins. Sugar cane and sugar beet based natural bio-polymers and thermoplastics possess wide applications in the use of structural composite materials, pharmaceuticals, electrical appliances, and electronic packaging materials. The production and thermal, mechanical (static and dynamic), and electronic characterization of these bio-polymers and their composites are discussed in detail in some investigations [1, 2]. The environmental sustainability, recyclability, and greenhouse gas (GHG) effects of these plastics and composites are found to be superior to those of epoxy and phenol based structural and electronic composites. The immense potential of vegetation based nano- and multi-scale composites with modifiers and reinforcements as the fillers in technological applications has been brought out by many investigators [3, 4].

Self-reinforced polymer composites (SRPC), which have the same material in different structures as the matrix and the reinforcement, possess comparable shear and tensile strengths unlike the glass, boron, or carbon fibre reinforced polymer composites. Due to their ultra-light weight specific properties (olefins like polyethylene and polypropylene are lighter than water), they are increasingly used in marine, transportation, electronics, and aerospace luggage applications. In this chapter, low- and high-density polyethylene, ultra-high-molecular-weight polyethylene (UHMWPE), polypropylene, polyamide, and other SRPC are evaluated for the purpose of their use in products and bio-disposal. As the same material is bonded to itself in as many structurally different forms as possible, a skilled tailoring of the micro-interface between the matrix and the reinforcement and

Acknowledgements: The authors immensely thank De Gruyter, Germany, for the opportunity to write the first and only book on self-reinforced composites. They also thank the management of VIT, Vellore, for all the encouragement and support.

https://doi.org/10.1515/9783110647334-010

its evaluation play a major role in deciding the macrostructural properties of the composite. Interfacial characterization using specially designed pull out tests and spectroscopy, static mechanical tests using computer controlled universal testing machines, drop mass impact tests, and post failure microscopic analysis reveal the relationship between interfacial properties and the static and dynamic mechanical properties of these ultra-light composite materials [5]. These composites have also been studied for their machining induced molecular changes that modify the mechanical properties and cause variations in damage [6]. Some of the benchmarked SRPC designs that find high-end applications in ballistic helmets and vests have been produced with novel manufacturing techniques. Current methods of improving the properties of matrix and reinforcement material, their processing, and modification of the interface for superior performance were discussed earlier in this book. Some of the SRPCs float on water when processed in any shape that provides for novel applications in air cargo systems and marine structures where floatability and light-weighting are important in energy and cost savings with safety included. As most of the SRPCs are thermoplastics, their recyclability, disposability, and degradability issues are taken into consideration. Carbon/carbon composites processed through the cost-effective polymer route find wide applications in tribology and ablation. Bio-friendly and bio-derived SRPCs which answer some of the environmental concerns are increasingly being designed and developed for similar applications not as lesser substitutes but as design and durability specific alternatives. This book concludes with projections on consumption, demand, and foreseen issues and challenges. Existing solutions and a vision for a consensual future for SRPCs are charted out.

1. Indian contribution: As India is a world class producer of sugar cane, sugar beet, other tubers like potato and vegetables with starch, cashew and badam, castor oil, and soybean, the quantum of bioplastics that can be produced from these conventional, organic and GM plants is immense. As on date, advanced and state of the art plastics and composites are being used in low-end applications as there is no incentive for farmers to produce plants and vegetables for the plastics and resins market exclusively. The use of advanced composites in low-end applications escalates costs and shifts the material consumption that would deplete the natural resources, through a mis-skewed usage at one end and a lack of demand for natural resources at the other. This invited chapter attempts to project the actual possibilities of the bio-resin and bio-plastic market in this country and provides the know-how for production of bio-polyethylene, bio-polylactic acid, bio-polyester, and castor oil based plasticizers. Their true potentialities in composites product applications involving structural, interior, electronic, and chemical engineering markets is discussed in this chapter. A novel working model with an economically feasible option is also provided for those concerned about their safe disposal, recycling, reuse, and conversion into useable fuel with little impact to the environment.

India is a world class producer of sugar cane, sugar beet, other tubers like potato and vegetables with starch, cashew and badam, castor oil, and soybean. The quantum of bio-resins and bio-plastics that can be produced from these conventional, organic, and GM plants is large.

India produces about 400 million tons of sugar cane per annum and all of it is used for the production of sugar. Though the technology for producing polyethylene from ethylene and ethanol distilled from sugar cane is available, the demand for sugar is so great that one has to look at non-edible and GM sugar cane production for the synthesis of polyethylene from sugar cane. The production of soybean can also be GM in order to assist in the production of less toxic epoxidized resins and not invade in to the edible sector. The farmers gain more from such GM cultivations as they generally yield more than the produce from conventional farming (say up to 94% as in GM crops). Bio-derived plastics are more resistant to pests and assist in the synthesis of more environment friendly resins and bio-plastics than their synthetic cousins. Besides, the fear and resistance factor offered to biotechnologically modified edible crop does not come in at all. The farmers, however, need an economical guarantee of demands for the inspiration. GM crops would also mean that they can be organically produced without any pesticides as they are basically resistant for any of the above applications. This approach improves economical production [1].

Table 10.1: Indian sugar cane production.

S. no.	States/UT	2020–21
1	Uttar Pradesh	177.67
2	Maharashtra	101.59
3	Karnataka	42.09
4	Tamil Nadu	12.80
5	Bihar	10.71
6	Gujarat	15.85
7	Haryana	8.53
8	Andhra Pradesh	4.12
9	Punjab	7.49
10	Uttarakhand	6.96
11	Madhya Pradesh	5.88
12	Telangana	1.36
13	West Bengal	1.56
14	Others	2.64
	All India	**399.25**

Table 10.2: Indian soybean production.

S.No.	States	Kharif 2021		
		Sowing area	Expected yield	Estimated production in million metric tonnes
1	Rajasthan	9.253	761	7.046
2	Madhya Pradesh	55.687	939	52.292
3	Maharashtra	43.848	1,102	48.325
4	Andhra Pradesh	–	NAN	
5	Chhattisgarh	0.513	910	0.467
6	Gujarat	2.237	1,015	2.271
7	Karnataka	3.827	1,005	3.846
8	Others	1.129	975	1.101
9	Telangana	3.488	1,015	3.54
	Grand Total	119.982	991	118.888

Tables 10.1 and 10.2 provide some information on India's superiority in the production of these crops. The Indian soybean production data stands at 12 to 13 million tons. A portion of it could be used in producing epoxidized soybean. It is observed that most of the plant oil production capabilities are confined to the southern states and Madhya Pradesh.

2. Global impact and solutions: Biopolyethylene (also known as renewable polyethylene) is polyethylene made out of ethanol, which becomes ethylene after a dehydration process. It can be made from various feed stocks including sugar cane, sugar beet, and wheat grain. One of the main environmental benefits of this process is the sequestration of roughly 2 kg of CO_2 per kg of polyethylene produced, which comes from the CO_2 absorbed by the sugar cane while growing, minus the CO_2 emitted through the production process. Over 1.5 billion pounds of CO_2 will be annually removed from the atmosphere. Dow and Toyota are making it happen. Polylactic acid is a transparent plastic produced from corn or dextrose. The biopolymer poly-3-hydroxybutyrate is a polyester produced by certain bacteria processing glucose, corn starch, or waste water. This is similar to polypropylene production.

Looking further ahead, some of the second generation bio-plastics manufacturing technologies under development employ the 'plant factory' model, using GM crops or GM bacteria to optimize efficiency. We know that castor oil derivatives are used as plasticizers in rigid plastics. Environmental fingerprinting and CO_2 emissions can be reduced by replacing petroleum derived plasticizers with castor oil plasticizers.

BASF – a German company is working on this aspect. This product is used in toys, impact-resistant plastics, hose pipes, medical aids, and so on.

Wood is a self-reinforced composite. As wood polymer composites can also be designed in a self-reinforced composition to minimize multi-material usage related costs, the technology status of wood plastic composites (WPCs) manufacture and processing is lucrative enough to venture in to such initiatives. WPCs are quite popular in construction industries. Their production uses a two-stage process that includes compounded pellets and shaping techniques. Commonly used processing includes sheets and profile extrusion, thermoforming, compression moulding, injection moulding, and a new trend in-line compounding and processing methods. The application benefits are (a) improved dimensional stability, (b) increased strength, (c) lower processing temperatures, (d) less energy usage, (e) increased heat deflection temperature, and (f) controlled thermal expansion. Up to 30% reduced cycle times are observed for injection moulded products where productivity is increased. The products also possess approximately 10–20% lower specific gravity in most of the cases and low volumetric costs.

As bamboo is a self-reinforced composite, an interesting product fabricated recently catches our attention. A bicycle has been fabricated by Prof. DP Mishra and his team from IIT Kanpur. This naturally derived product has been filed for patents and the entire product weighs just 8 kg. Figure 10.1 shows the complete product as it was designed and fabricated by his team [7].

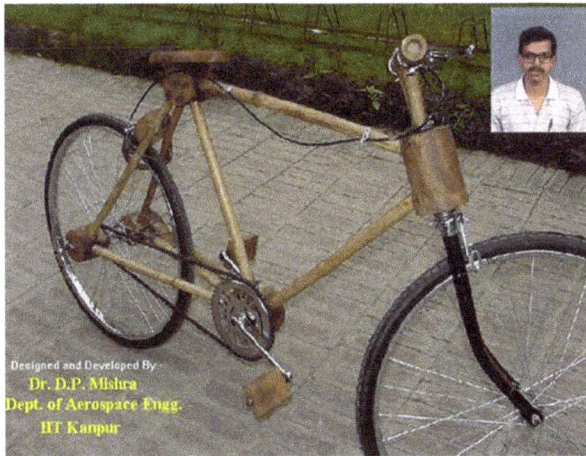

Figure 10.1: Picture of a bamboo bicycle as a wood-based SRC.

More products like polyethylene and polypropylene cargo containers that float on water in any shape are being manufactured in volumes. Imagine an entire aircraft with these cargo containers and baggage that meets with an accident. The survivors do not need a vest! Ballistic grade vests and helmets have also been manufactured from cross-plied polyethylene and UHMWPE materials and SRCs fabricated from polyethylene [8]. Figure 10.2 shows the author and team with a high-pressure mould-ing of an ultra-light ballistic grade helmet for the Indian Army [3]. Carbon/carbon nose cones and nozzles have been successfully produced as a national priority by a team from DRDL, Hyderabad, India [9]. Waste to wealth antimicrobial products and food packaging products have been developed using bio-polymers. The chicken breasts could be refrigerated under hygienic conditions when the polymer films were coated with sage extracts that were proven to be anti-microbial [10].

Figure 10.2: The author and team with a moulded polyethylene ballistic grade helmet.

3. Recycling waste to fuel and power generation: Cumulative plastics production in the world was 8 billion tonnes in 2018 and more than 60% of it was discarded. The oceans have been severely dumped with plastic wastes that have created a cli-mate change and pollution (Figure 10.3). Some of the beaches have been invaded by such plastic wastes resulting in tidal effects and a high GHG footprint (Figure 10.4). In Tamil Nadu, India, the use of plastic films below 30 micron thickness is banned. As recyclability is an issue, many other governments also regulate with suitable laws. In the case of thicker plastics, the process of recyclability is consid-ered economically viable. Many recycling projects can be found that recycle olefins up to a level that is economically feasible [11–13]. When a polymer is not recyclable, one may adopt a waste to fuel project through burning of olefins [14]. One of the foremost applications of these plastic and bio-plastic composites is their recyclabil-ity and ease of disposal through conversion of the bio-plastics in to useable fuel by depolymerization with the aid of a catalyst like zeolites, clay, and ammonium sul-phate and condensing the pyrolyzed gas in to fuel oil (Figure 10.5). Bio-plastic wax, kerosene diesel, and petrol could be obtained this way in an economical manner at affordable costs, where the fossil fuel is not depleted. Polyethylene and polypropylene wastes have been successfully converted into useable fuel this way. The final wastes can be incinerated using the waste-to-energy plants use household garbage including polymer composites as a fuel for generating power, much like other power stations use coal, oil, or natural gas (Figure 10.6). The burning of the waste heats water and the steam drives a turbine to generate electricity [12]. This knowledge chain sequence can be broken considering efficiency as a priority where the environmental impact would not be a priority. Environmental impact would be the least in a chain were recycling is followed by waste to fuel pyrolysis which in turn is followed by incineration to produce electricity and power. Here, the fossil fuels are not depleted due to a gradation of events that do not use a fresh batch of fossil fuels. But, if all the waste is incinerated at a first stage, an equal or more amount of fossil fuel would be required to supplement for regular plastics usage. However, this philosophy holds good only if all the three above said processes are carried out in a controlled manner where there are no toxin let-outs in to the atmosphere but usable sequestrations of by-products. Plastics that cannot be incinerated for power generation can be a treated with bacteria, as they tend to char during incineration and pollute the environment. Microbial digestion of polluting plastics is a known technology by now [15] though this option is not known to yield any calories.

Figure 10.3: Cargo from polyethylene and polypropylene would float after accidents.

Quantum of bio-plastics can be produced from the conventional, organic and GM plants. This chapter attempts to project the actual possibilities of the bio-plastic market in this country. Their true potentialities in composites product applications involving structural, interior, electronic, and chemical engineering markets is discussed. A novel working model with an economically feasible option is also provided

The turtle is attempting to eat a thin plastic sheet.

Figure 10.4: Cumulative plastics production in the world is 8 billion tonnes in 2018!!! And more than 60% is discarded!!!! The turtle is attempting to eat a thin plastic sheet.

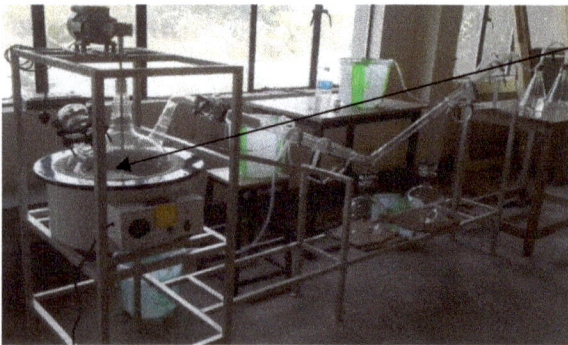

Glass chamber

Figure 10.5: Bio-plastic waste to fuel plant with pyrolyzing glass chamber and ammonium sulphate catalyst.

for those concerned about their safe disposal, recycling, reuse, conversion into useable fuel, and incineration for electricity production with no impact to the environment.

WASTE INCINERATION PLANTS
(SIMPLIFIED DIAGRAM)

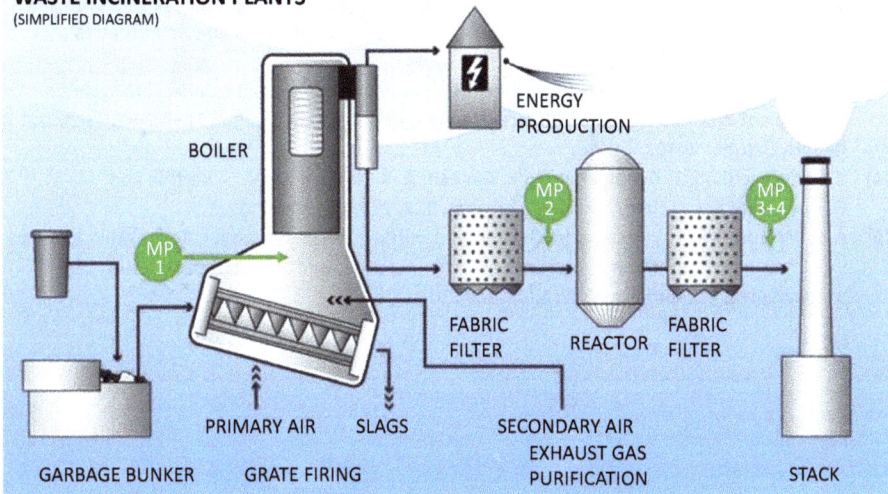

Figure 10.6: Incineration plant for power generation.

References

[1] K. Padmanabhan, Bio- resins and bio plastics synthesized from agricultural products for novel applications, Chapter No: 11, Biopolymers and Bio materials, APPLE Publishers, CRC, June 2018.

[2] A. Ajinkya Sawant and K. Padmanabhan, Synthesis and characterization of CNSL matrix compositions for composites applications, Chapter No: 4, APPLE Publishers, Biopolymers and Bio materials, CRC, June 2018.

[3] S. Harsha and K. Padmanabhan, Apple academic publishers, 2018.

[4] S. Harsha and K. Padmanabhan, Apple Academic Publishers, 2018.

[5] S. Chandran and P. Krishnan, "Preparation and characterization of self-reinforced fibre polymer composites with emphasis on the fibre/matrix interface", PhD Thesis, VIT, India, 2020. P. Krishnan as Guide.

[6] D. Akepati, P. Kuppan and P. Krishnan, "The influence of Novel machining induced molecular changes on the mechanical and damage characterization of self-reinforced polymer composites", 2021.

[7] Bamboo Bicycle, http://www.nitttrkol.ac.in/dpmishra/bamboo.php

[8] X. Chen, "Advanced Fibrous Composite Materials for Ballistic Protection, Woodhead Publishing Series in Composites Science and Engineering: Number 66." Amsterdam, Boston, Cambridge, 2016.

[9] R. Devi, DRDO products paper or patent details. http://www.drdo.gov.in

[10] N. Aziman, M. Jawaid, N. A. A. Mutalib, N. L. Yusof, A. H. Nadrah, U. K. Nazatul, V. V. Tverezovskiy, O. A. Tverezovskaya, H. Fouad, R. M. Braganca, P. W. Baker, S. Selbie and A. Ali, "Antimicrobial potential of plastic films incorporated with sage extract on chicken meat," *Foods*, 10, 2812, 2021, https://doi.org/10.3390/foods10112812.

[11] E. Kosior, J. Mitchell and I. Crezenzi, "Plastics recycling, January 2019," *Issues Environ. Sci. Technol.*, 47, 156–176, 2019.

[12] J. Hopewell, E. Kosior and R. Dvorak, "Plastics recycling: Challenges and opportunties'," *Philos. Trans. R. Soc. B*, 364, 1526, 27, July 2009.

[13] Factors affecting the life cycle assessment of biopolymers campbell skinner, life cycle assessment analyst, biocomposites centre, Bangor University, Bangor, LL57 2UW, Technical Report, October 2019.

[14] K. Padmanabhan, T. Deepak Kumar, P. Ganesh, A. V. Haricharan, R. S. Karthik and G. Devasagayam, *J. Energy Storage Convers.*, 3, 2, 225, July-Dec 2012.

[15] A. R. Mc Cormick, T. J. Hollein, M. G. London, J. Hittie, J. W. Scott and J. J. Kelly, "Microplastics in surface waters of urban rivers: Concentration, sources and associated bacterial assemblages," *EcoSphere*, 7, 11, 1, 08 Nov 2016.

Index

https://doi.org/10.1515/9783110647334-011

www.ingramcontent.com/pod-product-compliance
Lightning Source LLC
Chambersburg PA
CBHW081538220326
41598CB00036B/6474